EUGENICS SOCIETY SYMPOSIA

Volume 3

SOCIAL AND GENETIC INFLUENCES
ON LIFE AND DEATH

EUGENICS SOCIETY SYMPOSIA

Edited by
J. E. Meade and A. S. Parkes

1. Biological Aspects of Social Problems 1965
2. Genetic and Environmental Factors in Human Ability 1966

Edited by
Robert Platt and A. S. Parkes

3. Social and Genetic Influences on Life and Death 1967

SOCIAL AND GENETIC INFLUENCES ON LIFE AND DEATH

A Symposium held by the Eugenics Society in September 1966

Edited by

THE LORD PLATT

Chairman, Clinical Research Board,
Medical Research Council,
London, England

A. S. PARKES

Mary Marshall Professor of the Physiology of Reproduction,
University of Cambridge,
Cambridge, England

ℙ Springer Science+Business Media, LLC

ISBN 978-1-4899-6182-2   ISBN 978-1-4899-6365-9 (eBook)
DOI 10.1007/978-1-4899-6365-9

Library of Congress Catalog Card Number: 67–30833

First published 1967 by Oliver & Boyd, Ltd.
© Springer Science+Business Media New York 1967
Originally published by The Eugenics Society in 1967.
Softcover reprint of the hardcover 1st edition 1967

EDITORS' FOREWORD

THE EUGENICS SOCIETY's two previous symposia, *Biological Aspects of Social Problems* in 1964 and *Genetic and Environmental Factors in Human Ability* in 1965, have undoubtedly been successful in stimulating the minds of people who are interested in human affairs from many different points of view.

This volume brings to you the texts of the papers read at our Third Symposium, which was held in 1966, *Social and Genetic Influences on Life and Death*. Surely this is a broad enough title, a subject which could only be dealt with by taking a few examples of the important factors which determine the success or failure of human life from a biological point of view. So once again, we seek to bring to a wide public knowledge which might otherwise be available only in professional journals read by specialists. Once again, we took particular care in the choice of our speakers, and they probably never before have appeared together on the same platform. We are very grateful to them, not only for their spoken word but for letting us have their scripts for publication.

As before, we as Officers of the Eugenics Society take formal responsibility for the editing of the Volume; but we are only too willing to acknowledge that the real work has been done by Mrs K. Hodson, the editor of *The Eugenics Review*.

ROBERT PLATT
A. S. PARKES

CONTENTS

	PAGE
EDITORS' FOREWORD	v
GENERAL INTRODUCTION. LORD PLATT	ix
CONCEPTION, PREGNANCY AND BIRTH. *Chairman:* LORD PLATT	1
Occurrence and effect of human chromosome abnormalities. P. E. POLANI	3
Recent developments in medical genetics. C. A. CLARKE	20
Family growth and its effect on the relationship between obstetric factors and child functioning. RAYMOND ILLSLEY	29
Social and biological influences on foetal and infant deaths. THOMAS MCKEOWN	43
SOME MAJOR CAUSES OF ILLNESS: I. SOMATIC ILLNESS. *Chairman:* LORD PLATT	57
Recent trends in mortality. C. C. SPICER	59
Coronary thrombosis—progress and prospects. J. N. MORRIS	66
Environmental and constitutional factors in the aetiology of lung cancer. C. M. FLETCHER	78
SOME MAJOR CAUSES OF ILLNESS: II. PSYCHOLOGICAL ILLNESS. *Chairman:* SIR AUBREY LEWIS	93
Introduction. THE CHAIRMAN	95
Some observations on the prevalence of mental illness in contrasting communities. A. H. LEIGHTON	97
Social influences on the prevalence of psychiatric disorder. KENNETH RAWNSLEY	112
Genetic and social influences in the causation of suicide. ERWIN STENGEL	122
Alcoholism. NEIL KESSEL	130

PAGE

CAUSES AND EFFECTS OF AGEING.
     Chairman: PROFESSOR E. M. BACKETT   141

   Introduction.   THE CHAIRMAN   143

   Some economic and social consequences of ageing.   P. R. COX   145

   Psychiatric aspects of old age.   MARTIN ROTH   163

   Ageing and mortality.   R. D. CLARKE   183

   Genetic studies on longevity.   M. J. HOLLINGSWORTH   194

INDEX OF SUBJECTS   215

INDEX OF AUTHORS   218

GENERAL INTRODUCTION

SIR ROBERT PLATT
(now LORD PLATT)
Cambridge

UNFORTUNATELY Professor Parkes cannot act as Chairman at this first Session of our Symposium. This is a double disappointment to me, because I had intended to thank him for all that he has done in recent years for this Society. He was one, but only one, of the prime movers in starting these Symposia of the Eugenics Society of which this is the third. The first was on *Biological Aspects of Social Problems*, the second was on *Genetic and Environmental Factors in Human Ability* and this one is on *Social and Genetic Influences on Life and Death*.

Needless to say, this has a rather more medical flavour than the other two, but is not too specialized for intelligent lay people to be able to understand it and also be deeply interested in it. The theme running through these Symposia is the feeling of this Society that perhaps its main *raison d'être* nowadays is that it brings together people who approach human problems from different aspects—sociologists, geneticists, medical people, educationalists and many others.

In our sessions we are only, of course, touching upon a sample of the various circumstances which influence human life, the quality of human life and the future of the human race from birth to death. What comes out of these discussions should very gradually, in the course of time, rouse more and more people to become interested in these problems. One hopes that this will eventually be to the betterment of the human race, which brings in the ' eugenic ' part of the title of this Society.

CONCEPTION, PREGNANCY
AND BIRTH
Chairman: LORD PLATT

OCCURRENCE AND EFFECT OF HUMAN CHROMOSOME ABNORMALITIES

PAUL E. POLANI

Guy's Hospital Medical School, London

CHROMOSOME anomalies in man are relatively common changes of the genetic material, very often lethal to the conceptus before birth. Among the survivors, they generally cause a range of developmental deviations, from severe somatic anomalies with grave central nervous system involvement and a shortened life to less striking somatic changes, or sometimes to minimal changes only, but with infertility or sub-fertility and with mental retardation. In some instances, the changes may be minimal and subtle and cause little more than an impairment of psycho-social adaptation, resulting in deviant behaviour. To a large extent, the different outcomes are related to the different types of chromosome abnormalities and to the specific chromosomes involved in them.

From what has been learned so far, certain generalizations are possible:

Triploidy, the condition in which the basic chromosome set of 23 is present in the cells three times instead of the normal twice of diploids, the chromosome number being 69 instead of 46, is incompatible with post-natal existence. By contrast, trisomy, the presence of an extra chromosome, with a chromosome number of 47, appears to be less injurious and therefore is compatible with survival at birth. Certain distinctions must, however, be made: (1) Of the non-sex chromosome trisomies—the autosome trisomies—only three have been found with a certain frequency among survivors, trisomy of number 18, of D_I (probably number 13) and of number 21, the latter the cause of Down's syndrome or mongolism. But in these there is an interesting pattern of post-natal death. The average age at death of children with trisomy for number 21 (Down's syndrome) is just over ten years,[5] that of D_I trisomy (Patau's

3

syndrome) just under three and a half months, and that of trisomy for number 18 (Edwards's syndrome) just over two and a half months.[25] The rest of the autosome trisomies appear to be nearly always prenatally lethal. (2) By contrast, sex-chromosome trisomies would appear only seldom to cause prenatal death, and to cause far less serious abnormality in survivors than the autosome trisomies do. Further evidence of the greater tolerance of man to sex-chromosome imbalance is the fact that higher polysomy of the sex chromosomes has been found in survivors. Thus several males have been described with as many as four X chromosomes and a Y, and even a not very abnormal female with five X's has been detected, with a chromosome complement of 49.

While trisomy, particularly of the sex chromosomes, but also of some of the autosomes, is compatible with survival at birth, the position as regards monosomy is very different. Monosomy is the absence of one chromosome from the proper diploid complement. The sex chromosomes are affected by the only fully monosomic condition, when, instead of being paired with an X or a Y, a single X chromosome is present, thus giving a chromosome complement of 45. Here again we have evidence of a considerable tolerance of man to sex-chromosome imbalance, but this is only relative, as there is ground for believing that this condition, the so-called XO state, is very often prenatally fatal. In addition, many of the affected females surviving at birth are severely malformed, having what is named ' Turner's syndrome ', and practically all are sterile. Related to monosomy are chromosome deficiencies. Quite a rare one affects one of the autosomes of the group 4–5, the B group. It affects only one part of one member of a pair of chromosomes of this group, to be precise only the tip of the short arm of one member of the relevant chromosome pair. It produces a syndrome of mental retardation with a distressing and cat-like cry in infancy known as the Cri-du-Chat syndrome.[25] A second, and still rarer, type of deficiency, of which a dozen or so cases have been seen, involves the tip of one chromosome of pair number 18—another structural chromosome error with deficiency—and is often responsible for multiple somatic anomalies. The third type of deficiency has been described in a few persons.[11, 18, 21] It involves chromosome 21, the chromosome which, when trisomic, is the cause of mongolism. In the deficiencies of this chromosome, the morphological traits of the patients were interpreted as being rather the opposite of the traits in Down's syndrome, and the collective name of ' anti-mongolism ' has been used to describe them. The clinical abnormality in this case can be the result of structural anomaly with deficiency as well as chromosome mosaicism, in which some cells have the right chromosome number while others lack a chromosome of the G group, have 45 chromosomes, and are interpreted as monosomic for chromosome number 21. By chromosome mosaicism is meant the presence side by side in the same organism of two, or sometimes more, cell lines with different chromosome complements. Many make a distinction between mosaicism and chimaerism and stipulate that in mosaicism, apart from the differences in chromosome make-up, the basic genetic constitution must be the same in the different types of cells. This happens when the cell lines arise from an error of early cleavage in a zygote, the result of which is developmental chromosome mosaicism. In chimaerism, on the other hand, the implication is generally that the two cell lines are from different zygotes with different genetic make-ups and that they happen to be in one individual as a result of an accident of nature or an experiment of man.

Biologically, there are some very important general consequences of developmental chromosome mosaicism.

Firstly, the error that causes it is an error of chromosome partitioning at a somatic, or mitotic, cell division of a normal or abnormal zygote, and occurs early in its life history.

Secondly, each cell line tends to make its own separate contribution to the total phenotype, which is therefore of intermediate or possibly of patchy type. If, for instance, there are two cell lines, one with 46 chromosomes and normal, the other with 47 and with, for example, three chromosomes number 21, the mosaic subject may well have features of mongolism, but there will be a dilution of the abnormal phenotype. The physical signs of Down's syndrome will therefore be less obvious ; indeed, in some instances of this type of mosaicism only micro-signs of the condition may be found and may require a careful search, and the intellectual performance of the subject may be considerably nearer normal.

The third general consequence of developmental chromosome mosaicism is that it can involve a greater or smaller proportion of germ cells. Since that is so, the abnormal germ cells may produce abnormal gametes and the carrier of mosaicism may, therefore, produce abnormal children, unless this is prevented from happening because of oddities of chromosome segregation or because of selection at gametic level. This tendency of mosaics to produce abnormal children has been verified in Down's syndrome, in which it was high-lighted by the fact that in some of the reported cases the chromosome-mosaic mothers were phenotypically almost normal.[3, 7, 23, 29, 31] They were in fact identified, not because of their own features, but because they had produced trisomic children with Down's syndrome.

A fourth interesting event which ties up with mosaicism is the formation of uni-ovular discordant twins, for instance same sexed but discordant for Down's syndrome,[12] or even different sexed, though uni-ovular, discordant for Turner's syndrome, the female being XO and affected and the uni-ovular co-twin a normal XY male.[27] The explanation is simply that the error of mitosis which results in the chromosome abnormality is coupled at the same time with the separation from each other of the blastomeres which carry the products of faulty chromosome partitioning, with the result that chromosomally dissimilar twins are formed.

Finally, a rare but interesting outcome—which is 'chromosome mosaicism' in a general sense and yet at times can be demonstrated to be a form of chimaerism in the narrow definition of the word—is the XX/XY person. The affected person may have true hermaphroditism, i.e. may carry both ovarian and testicular tissue.[1, 10, 15, 17, 22]   Incidentally, true hermaphrodites with this constitution are about 7·5 per cent of all chromosomally studied true intersexes.

One of the likely events that can produce an XX/XY individual, and for which there is indirect confirmation in man, is double fertilization by an X- and a Y-bearing sperm. In this event, in addition to the two paternal sex chromosomes, two paternal autosomal contributions should be detectable in suitable circumstances, if a study is made of genetic marker characters, such as autosomal blood groups, serum proteins and enzymes, etc. The maternal genetic contribution may also be double; in which event there are two possible interpretations of the origin of the chimaerism, namely fertilization of two ova by two sperm and their fusion into one zygote, or double fertilization of both ovum and polar body. When there is only a single maternal genetic contribution, a possible interpretation that has been put forward is that, following penetration of the ovum by two sperm, its haploid complement undergoes replication so that each sperm can unite with one haploid set of maternal origin, the two sets being genetically identical. Naturally the detectability of the two possible parental contributions depends on the parents being heterozygous for genes whose alleles determine distinct characters. In some cases, the number of such genes may be too small to be helpful in rejecting with confidence the possibility of a double genetic contribution from that parent. The possibility that an XX/XY mosaic may result from the fusion of two normal fertilized ova— a sort of conjoining of dizygotic twins—does not meet with general support. Tarkowski,[24] who, like others, has obtained this type of fusion experimentally in mice, after, for instance, denudation of the cleaving eggs, does not think that this could happen *in vivo*. One of the other mechanisms for the origin of XX/XY persons may therefore be more likely.

It is not surprising that an XX/XY person may often be a true hermaphrodite with ovarian and testicular tissue, but it is interesting that confirmation of this is available from the results of experiments in mice involving *in vitro* fusion of genetically different ova, their cultivation in the fused state to the stage of a single blastocyst, followed by implantation of the chimaeric blastocyst into foster mothers.[14, 24] In these circumstances one would expect, on an average, half the chimaeric blastocysts to be XX/XY and the rest to be equally distributed between XX females and XY males. Naturally also, in man double fertilization chimaeras will be wholly XX or XY and should arise at random with equal combined frequency as the XX/XYs. It is interesting that neither in experimental chimaerism in the

B

mouse, nor in the natural XX/XY condition in man, are all subjects with this chromosome complement true hermaphrodites. In man some XX/XY persons have a fully (or nearly so) male phenotype, including testes; others have a female phenotype.[1] In the mouse, out of sixteen chimaeras in which the sex phenotype was investigated,[24] only three were hermaphrodites; of the remaining thirteen, eleven were male and two female. Although the numbers are small, the explanation given of these findings is that a proportion of XX/XY mice develop into phenotypic males, indistinguishable from normal.

The experimental work in mice is also interesting in relation to chromosome mosaicism. It is clear that developmental chromosome mosaicism, if it is to be of sufficient importance to alter embryonic development, must arise during early division of a primarily normal, or abnormal, zygote. If the mitotic error which causes mosaicism with two cell lines occurs early, it may be pertinent to ask whether these lines may not become separated between embryo proper and trophoblast,[2] as a result of what has been termed an anti-mosaic effect of mammalian eggs. Such a separation should have far-reaching consequences on the developing embryo. Some of the work [16] can be taken to support the possibility of an early segregation of blastomeres with different destinies. Other work, for instance the experimental fusion of denuded cleaving eggs, sometimes many at a time including some with tagged cells, has shown the mouse egg to be 'morphogenetically labile'.[14] For this animal at least, it would appear difficult to imagine a polarity of segregation of chromosomally different blastomeres, though perhaps the matter is not settled.

It might be appropriate now to consider the frequencies of some of the more common anomalies, mainly of number, that have been mentioned. Estimates of some of these anomalies in clinically selected populations have been available for some years, following the introduction of the simple nuclear-sexing techniques capable of indirectly demonstrating in cells the presence of one or two X chromosomes. As is well known, normal males are chromatin negative by this method, because they have a single X, while normal females, with two X's, are chromatin positive. The frequency of abnormal chromatin positive males in the educationally subnormal, in the mentally defective and among infertile or subfertile males with Klinefelter's syndrome attending special clinics had been worked out even before chromosome studies had indicated that these males either had an XXY sex-chromosome complement or were sex-chromosome mosaics, mostly XY/XXY. The frequency of abnormal chromatin negativity among women with simple ovarian dysgenesis and with Turner's syndrome had also been estimated before it was proved that most of them had an XO sex-chromosome complement. Only later was it shown, with the aid of chromosome studies, that chromatin-positive women with ovarian dysgenesis either had X-chromosome mosaicism or harboured structurally abnormal X chromosomes with or without mosaicism.

Since 1959, the population frequency of chromosome abnormalities has received special attention through a number of studies on newborn infants done in maternity hospitals. These have for the most part involved routine nuclear-sexing studies, usually coupled with chromosome investigation when discrepancies between phenotype and sex-chromatin findings were demonstrated. In all, well over 40,000 infants, both males and females, have been studied to date (summarized, with additional data, by Taylor and Moores).[26] In addition, studies for the detection of autosome anomalies have been done on clinically selected patients. Older clinical estimates of the frequency of Down's syndrome among the newborn are also available. From all these sources, the overall frequency of chromosome anomalies among newborn infants of both sexes can be set at about 1/250 (or about 0·40 per cent). Meanwhile, work has been done on series of spontaneous and induced abortions, in order to arrive at an estimate of the frequency of chromosome anomalies as a cause of spontaneous abortion and to get nearer the incidence of chromosome anomalies at conception. For various and obvious reasons it is important to assess the load of human chromosome mutation, preferably in different populations, and in one population at different times and in varying circumstances. This work was reviewed last year[20] and more recently at a meeting of a scientific group at the World Health Organization in Geneva.[32] Most of the figures quoted are from the data collected for that meeting and include up-to-date information by various workers in this field. The abortion material was submitted to nuclear-sexing study and to culture for chromosome investigation. Apparently no selection was made, though an attempt was made to classify the abortions into spontaneous and induced, and a description of the abortus was usually provided. About half of all spontaneously aborted conceptions failed to yield chromosome results. Of those that did, about one in five was chromosomally abnormal and there was a wide range of anomalies (Table I). The proportions of the different anomalies among all chromosomally abnormal spontaneous abortions are indicated in the first line of the Table, and those among all spontaneous abortions in the second; trisomies are about 41 per cent of all anomalies and it can be assumed that the great majority of them are of the autosomes. The possibility is not excluded, however, that among the C-group trisomics there might be the odd one with three X chromosomes, but nuclear-sexing studies indicate that this could only rarely be so. Equally possible, an occasional G chromosome might be a Y, though this chromosome is usually easily identified and can be distinguished from the G autosomes, numbers 21 and 22. Among the remaining anomalies—over half of the total—the single largest group is represented by XO conceptuses. It is possible that an odd abortus classified as XO might be monosomic for an autosome of the C group, but again the sex-chromatin studies carried out concurrently indicate this to be unlikely. Also striking is the high proportion of triploids, with 69 chromosomes, second in frequency only to the XOs. In triploids, the sex-chromosome complement could be either XXX or XXY, if it is always the mothers that contribute two haploid sets. If, however, the double contribution were, in some cases at least, paternal, then not only XXX and XXY but also XYY triploids should result. Of fourteen triploid spontaneous abortions with a classifiable sex-chromosome complement, four were XXX, one was XYY and nine were XXY.[20] The numbers are exiguous, but if this trend were to continue it would suggest that, in man, errors in the ovum contribute to triploidy more often than those in the sperm. Among the autosome trisomies,

TABLE I

DISTRIBUTION OF CHROMOSOME FINDINGS IN 153 (19%) CHROMOSOMALLY ABNORMAL OF 788 SPONTANEOUS ABORTIONS. (W.H.O. Memorandum [32]; from the literature and personal communications.)

| AUTOSOME TRISOMIES, 41% | | | | | OTHERS, 59% | | | |
A–C (1–12, X)	D (13–15)	E* (16–18)	F (19–20)	G (21–22, Y)	XO	Triploids	Others	
8	6	14	·65	12	21	17	21	% of abnormal (99·65)
1·6	1·1	2·7	·13	2·4	4·0	3·3	4·0	% of all (19·2)

* Mostly number 16

those of the F group are surprisingly rare, and those of the E group rather common. Among the latter, however, trisomy 16, which involves a chromosome easily identifiable in a positive manner, outweighs the other two possible trisomies of the group by three to one.[20]

In the aggregate of series under discussion, about two-thirds of the spontaneous abortions that were dated occurred in the first trimester and one-third in the second. There is an interesting, though not surprising, distribution of the frequency of anomalies in the two groups. Four-fifths of the spontaneous abortions in the first trimester were chromosomally abnormal, as against one out of every five of the losses in the second trimester. Not unexpected is the fact that chromosome abnormality was about nine times more common among the morphologically abnormal conceptuses than among those that looked superficially normal, the percentages of anomalies being 61 and 7 per cent respectively.

These figures can be used to obtain an approximation to the contribution that pregnancies terminating in spontaneous abortion make to the incidence of chromosome anomalies in recognizable pregnancies, though obviously the ultimate aim is the frequency of chromosome anomalies at conception. From this estimate an approximate calculation can be made of the proportion of pre-natal loss of certain specific chromosome anomalies, and therefore the strength of pre-natal selection against them. Taking not only those losses that occur in the first three or four months, 15 per cent would appear to be a reasonable overall estimate of the frequency with which known pregnancies end in a spontaneous abortion.[30] At this level, the overall contribution of abortions to the frequency of chromosome abnormalities in recognizable pregnancies is 2·88 per cent (Table II). Triploids are a little more than one-quarter,

TABLE II

RECALCULATION OF DATA OF TABLE I, ASSUMING THAT SPONTANEOUS ABORTIONS ARE 15% OF ALL RELEVANT PREGNANCIES
(Expressed as frequencies per cent)

AUTOSOME TRISOMICS 1·18%			OTHERS 1·7%			ALL
A–C E–F	D	G	XO	Triploids	Others	1/35
·66	·16	·36	·60	·50	·60	2·88

and sex-chromosome anomalies—namely XOs—perhaps one-third, or a little more, of the autosome anomalies. A comparison of these figures with those on survivors at birth (Table III)

TABLE III

FREQUENCY OF CHROMOSOME ANOMALIES AT BIRTH
(Data from literature and additional data.[26])   (Percentage of liveborn of both sexes.)

AUTOSOME TRISOMICS			SEX-CHROMOSOME TRISOMICS		OTHERS		ALL
D_1	18	21	XXX	XXY, etc.	XO	Others	1/244
·022	·015	·152	·034	·101	·018	·068	·41

shows that, among the latter, sex-chromosome and autosome anomalies are in approximately equal numbers; but this is due to a high proportion of sex-chromosome trisomics, while the proportion of XOs for newborn infants of both sexes is only about one in 5500. The overall frequency of chromosome anomalies in all conceptions that implant, grow and result in a recognizable pregnancy (Table IV) is estimated at 3·3 per cent with a pre-natal lethality as spontaneous abortions of 88 per cent. Thus pre-natal selection against these unbalanced chromosome complements appears to work very effectively, and continues to do so in the survivors at birth, to judge by their higher mortality and their diminished fertility. Maintenance of these chromosome anomalies in the population is, therefore, by and large, the result of repeated mutational events related in part to parental (particularly maternal) ageing, possibly to genotypic effects, and to other events whose nature eludes us to a greater or lesser extent. Different anomalies vary in respect to pre-natal death. For instance, XOs might be 0·62 per cent of all conceptions, with a pre-natal mortality as spontaneous abortions of about 97 per cent. Full triploids, with a frequency of at least 0·5 per cent at conception (granted the assumptions made above), are apparently not found among survivors at birth, though three or four triploid/diploid mosaics have been described.[4, 6, 8, 9] Full triploids would appear to have a 100 per cent pre-natal mortality. Autosome monosomics are found neither among

TABLE IV

FREQUENCY OF CHROMOSOME ANOMALIES (PER CENT, BOTH SEXES) IN ALL
CONCEPTIONS THAT RESULT IN A RECOGNIZABLE PREGNANCY (first row),
AND IN SURVIVORS AT BIRTH (second row, from Table III).
(It is assumed that 15% of all pregnancies terminate in
spontaneous abortion (see text).)

CHROMOSOME ANOMALIES FREQUENCY %	AUTOSOME TRISOMIES:		SEX-CHROMOSOME ANOMALIES		TRIPLOIDS	OTHERS	ALL
	A–F (1–20)	G (21–22)	Excluding XO	XO			
Estimate in all conceptions that implant and grow. (A)	·857	·512	·163	·618	·500	·640	3·29
In survivors at birth	·037	·152*	·163	·018	? nil	·040	·41
Estimated pre-natal loss (%) of (A) as abortion		70%	? nil	97%	100%		88%

* Down's syndrome

survivors at birth nor among spontaneous abortions. They
should be quite frequent among conceptions, but might cause
pre-implantation, or very early post-implantation, loss, too
early to be clearly noticed as abortions or for the conceptus to
be detectable clinically. The case of G-autosome trisomics
requires special mention. It is generally believed that trisomy
of a specific chromosome of this group, by definition number
21, is the cause of Down's syndrome, or mongolism. A definite
syndrome corresponding to proved trisomy of the other autosome of the group, number 22, has not been identified,[19]
and of the few possible cases described with this presumptive
chromosome complement some have been reviewed and shown
to carry in fact an extra Y chromosome [28] and not a G autosome as originally suspected. The problem of trisomy of
number 22 is unresolved, and unfortunately auto-radiography,
which can help to identify other chromosomes, is not very
useful in this instance because of the small size of the autosomes

in question. There is, therefore, the possibility that the G-autosome trisomic abortuses are trisomic for number 22. However, if we were to make the assumption that it is number 21, then the overall frequency of trisomy of number 21, and therefore potentially of Down's syndrome, would be about 0·5 per cent of all recognizable pregnancies with a pre-natal mortality of 70 per cent.

Whatever the real nature of the extra G autosome, and whatever its relationship to Down's syndrome, it can be estimated that the frequency of non-disjunction (i.e. failure of the normal separation of chromosomes of the same pairs) in the G-autosome group, for instance during meiosis, is about one per cent. This average frequency could all be attributed to female meiosis in view of the maternal-age effect on the incidence of Down's syndrome, namely of trisomy of number 21, and apparently also on trisomy G in spontaneous abortions.[20] But, going further and relating the data to different maternal ages indicates that non-disjunction of a G autosome, leading in half the cases to di-somic gametes and hence to trisomic zygotes, and in the other half to nulli-somic gametes and possibly, therefore, to monosomic zygotes, could occur in about one in every six maturing ova at the mean maternal age of 47½ years, to take one example. This is obviously remarkable, and equally important is the fact that some of the other chromosome anomalies also show a maternal-age effect, and consequently that their mechanism of origin could be similar to that which leads to non-disjunction of an autosome of the G group. Because of this striking maternal-age effect in Down's syndrome, and also in the other trisomies, both of autosomes and sex chromosomes, it has been calculated by Lenz[13] that a reduction of five years in the child-bearing age after the age of thirty could decrease the incidence of trisomic children by some 40 per cent.

At any rate there is this striking relationship between maternal age and chromosome non-disjunction. Its nature is unknown and its cause remains to be elucidated.

DISCUSSION

Asked if he could give an estimate of the proportion of Down's syndrome in regular trisomy 21 which was due to mitotic rather than meiotic non-disjunction, PROFESSOR POLANI said that some idea could be derived from the data on the proportion of children with mongolism who had clear-cut mosaicism. There was, of course, the question of clinical selection, and there were other problems, but in children with mongolism who had been referred with a firm diagnosis, the proportion of *detectable* mosaicism was about 2 or 3 per cent; in children referred with a doubtful diagnosis, the proportion rose rapidly to 10 or 12 per cent.

Questioned as to whether he could expand his earlier suggestion that abnormal psycho-social behaviour was related to abnormal karyotypes, Professor Polani said that this appeared to be the case among individuals with sex chromosome abnormalities. Males with the Klinefelter syndrome, who were mostly XXY—with 47 chromosomes—had been known to show deviant behaviour; some of the females with XXX chromosomes also appeared to be in this category. Here there was need for further study. More recently it had been emphasized that males with YY chromosomes, with X and with XX, had difficulty in social adaptation as well as being tall and aggressive, which made them rather formidable. The data were based on surveys from special institutions for criminal defectives; there the incidence of XYY was about 3 per cent, whereas in the general population it was, roughly, about one in 2000.

A questioner asked whether there was any possibility of *eugenic* effects resulting from chromosome abnormalities. Professor Polani said that he knew of no evidence that such abnormalities as had been detected in man could be of eugenic value. It was in the nature of investigations of this type that the highest yield was from abnormal rather than from normal people, but studies of normal people had shown *minor* variants in karyotype with a certain frequency. This, too, was a field which merited further study. Another speaker pointed out that there must have been one eugenic chromosome interchange

—the reduction of the chromosomes from 48 in the gorilla and chimpanzee to 46 in man. Professor Polani replied that it had been suggested that if a variant of human being were to be developed with D/G interchanges, homozygous for interchanges between the 21 and one D chromosome, this type would not produce mongol babies. But this was a very slim argument.

The discussion then turned to the question of a relationship between paternal age and Down's syndrome. Professor Polani said that Professor Penrose had published a paper in the *Lancet* in which he reported a possible paternal-age effect in cases where there had been an interchange of the G/G type—the 21 : 22 or 21 : 21 type. Further data had not fully substantiated this, but that only meant that there might be two populations in this group, one in which there was a paternal-age effect and another in which there was not. This was quite logical if one considered that cytological mechanisms might underlie the presumptive G/G interchange. Again, paternal age seemed to increase other chromosomal abnormalities, particularly the so-called iso-chromosomes for the long arm of the X, with mosaicism; in these there was a clear-cut, though not very strong, paternal age effect which might also occur in one or two other chromosome abnormalities. It was to be noted that there was no parental age effect in the XO condition, either among survivors or among abortions; this suggested that the mechanism of origin was quite different from that which one associated with advancing maternal age and Down's syndrome.

REFERENCES

(The references quoted are mostly reviews or papers surveying the literature)

1. BAIN, A. D. and SCOTT, J. S. 1965. Mixed gonadal dysgenesis with XX/XY mosaicism. The evidence for the occurrence of fertilization by two spermatozoa in man. *Lancet* i, 1035.
2. BEATTY, R. A. 1957. Parthenogenesis and polyploidy in mammalian development. *Cambridge Monographs in Experimental Biology No. 7* (p. 82). Cambridge University Press.
3. BLANK, C. E., GEMMELL, E., CASEY, M.D. and LORD, M. 1962. Mosaicism in a mother with a mongol child. *Brit. med. J.* ii, 378.
4. BÖÖK, J. A., MASTERSON, J. G. and SANTESSON, B. 1962. Malformation syndrome associated with triploidy—further chromosome studies

of the patient and his family. *Acta genet. statist. med. (Basel)* **12**, 193.

5. BROTHERS, C. R. D. and JAGO, G. C. 1954. Report on the longevity and the causes of death in mongoloidism in the state of Victoria. *J. ment. Sci.* **100**, 580.

6. ELLIS, J. R., MARSHALL, R., NORMAND, I. C. S. and PENROSE, L. S. 1963. A girl with triploid cells. *Nature, Lond.* **198**, 411.

7. FERRIER, S. 1964. Enfant mongolien — parent mosaïque. Étude de deux familles. *J. Génét. hum.* **13**, 315.

8. FERRIER, P., FERRIER, S., STALDER, G., BÜHLER, E., BAMATTER, F. and KLEIN, D. 1964. Congenital asymmetry associated with diploid-triploid mosaicism and large satellites. *Lancet* i, 80.

9. GROPP, A., JUSSEN, A. and ODUNJO, F. 1964. Near-triploid chromosome constitution in epithelial-cell cultures of palatal mucosa from a case of cleft palate. *Lancet* i, 1167.

10. KLEMPMAN, S. and WILTON, E. 1963. True hermaphroditism: chromosomal analysis. *S. Afr. med. J.* **37**, 1096.

11. LEJEUNE, J., BERGER, R., RÉTHORÉ, M.-O., ARCHAMBAULT, L., JÉRÔME, H., THIEFFRY, S., AICARDI, J., BROYER, M., LAFOURCADE, J., CRUVEILLER, J. and TURPIN, R. 1964. Monosomie partielle pour un petit acrocentrique. *C.R. Acad. Sci. (Paris)* **259**, 4187.

12. LEJEUNE, J., LAFOURCADE, J., SCHARER, K., DE WOLFF, E., SALMON, C., HAINES, M. and TURPIN, R. 1962. Monozygotisme hétérocaryote, jumeau normal et jumeau trisomique 21. *C.R. Acad. Sci. (Paris)* **254**, 4404.

13. LENZ, W. Prospects in prevention of trisomic conditions based on maternal age. Lecture to the Joseph P. Kennedy, Jr., Foundation. The *Third International Scientific Symposium on Mental Retardation*, 11th April, 1966, at Boston, Mass. To be published.

14. MINTZ, B. 1965. Nucleic acid and protein synthesis in the developing mouse embryo. In *Pre-implantation Stages of Pregnancy* (p. 145). *Ciba Foundation Symposium* (Ed. G. E. W. Wolstenholme and M. O'Connor). London, Churchill.

15. MONTERO, E., SALVATIERRA, V., GUIRÃO, M. and SASTRE, M. 1965. Análisis citogenético de un hermafrodita verdadero alternante. *An Dessarrollo* **13**, 277.

16. MULNARD, J. G. 1965. Studies of regulation of mouse ova *in vitro*. In *Pre-implantation Stages of Pregnancy* (p. 123). *Ciba Foundation Symposium* (Ed. G. E. W. Wolstenholme and M. O'Connor). London, Churchill.

17. OVERZIER, C. 1964. Ein XX/XY-Hermaphrodit mit einem 'intratubulären Ei' und einem Gonadoblastom (Gonocytom III). *Klin. Wschr.* **42**, 1052.

18. PENROSE, L. S. 1966. Anti-mongolism. *Lancet* i, 497.

19. POLANI, P. E. 1964. Chromosome anomalies. *Ann. Rev. Med.* **15**, 93.

20. POLANI, P. E. 1966. Chromosome anomalies and abortions. *Devl. Med. child Neurol.* **8**, 67.

21. REISMAN, L. E., KASAHARA, S., CHUNG, C.-Y., DARNELL, A. and HALL, B. 1966. Anti-mongolism. Studies in an infant with a partial monosomy of the 21 chromosome. *Lancet* i, 394. (See also: REISMAN, L. E. 1966. Anti-mongolism. *Lancet* i, 602 and *Lancet* ii, 57.)

22. SEGNI, G. and GROSSI-BIANCHI, M. L. 1965. Un raro caso di ermafroditismo vero di tipo alterno con cromosomi XX/XY. *Minerva paediat.* 17, 983.

23. SMITH. D. W., THERMAN, E. M., PATAU, K. A. and INHORN, S. L. 1962. Mosaicism in mother of two mongoloids. *Amer. J. Dis. Child.* 104, 534.

24. TARKOWSKI, A. K. 1965. Embryonic and postnatal development of mouse chimeras. In *Pre-implantation Stages of Pregnancy* (p. 183). *Ciba Foundation Symposium.* (Ed. G. E. W. Wolstenholme and M. O'Connor.) London, Churchill.

25. TAYLOR, A. 1966. Patau's, Edwards' and Cri-du-Chat syndromes. A tabulated summary of current findings. *Devl. Med. child Neurol.* 9, 78.

26. TAYLOR, A. and MOORES, E. *Sex-chromatin Survey of Newborn Children in Two London Hospitals.* [To be publ.]

27. TURPIN, R., LEJEUNE, J., LAFOURCADE, J., CHIGOT, P.-L. and SALMON, C. 1961. Présomption de monozygotisme en dépit d'un dimorphisme sexuel : sujet masculin XY et sujet neutre haplo X. *C.R. Acad. Sci. (Paris)* 252, 2945.

28. UCHIDA, I. A., MILLER, J. R. and SOLTAN, H. C. 1964. Dermatoglyphics associated with the XXYY chromosome complement. *Amer. J. hum. Genet.* 16, 284.

29. VERRESEN, H., VAN DEN BERGHE, H. and CREEMERS, J. 1964. Mosaic trisomy in phenotypically normal mother of mongol. *Lancet* i, 526.

30. WARBURTON, D. and FRASER, F. C. 1964. Spontaneous abortion risks in man : Data from reproductive histories collected in a medical genetics unit. *Amer. J. hum. Genet.* 16, 1.

31. WEINSTEIN, E. D. and WARKANY, J. 1963. Maternal mosaicism and Down's syndrome (mongolism). *J. Pediat.* 63, 599.

32. (World Health Organization) Geneva Conference. 1966. Standardization of procedures for chromosome studies in abortion. *Bull. Wld. Hlth. Org.* 34, 765.

RECENT DEVELOPMENTS IN
MEDICAL GENETICS

C. A. CLARKE

Department of Medicine, University of Liverpool

I HAVE been asked to discuss recent developments in medical genetics in relation to conception, pregnancy and birth and this is a tall order to fulfil in twenty-five minutes. Fortunately Professor Polani has already described one big aspect, namely the way in which chromosomal abnormalities affect this interesting period in our lives, and it seems to me that a second big field in which there have been exciting advances is immunology. For example, the reasons why a tissue is rejected are being increasingly understood, and successful organ transplantation seems just round the corner—though on the other hand, why the foetus is not rejected is still something of a mystery. Then there are the disorders resulting from blood group incompatibility, of which the Rhesus situation is much the most important, and one which seemed to me to be highly relevant to the general subject of this symposium. Furthermore, I have a particular interest in it, and I am going to stick to this one point and tell you how I think Rh haemolytic disease may be prevented.

The date I begin with is 1943,[6] when Levine showed that ABO incompatibility as between mother and foetus was nearly always protective against Rh immunization (Figure 1), i.e. there was in existence a naturally occurring mechanism which prevented Rh haemolytic disease in a particular type of case. About 1957, we were vaguely wondering if we might be able to mimic this protection in the far more frequent situation where the mother and foetus are compatible on the ABO system. We went on wondering until 1959, when Zipursky[9] used the Kleihauer elution technique [5] to demonstrate foetal red cells in the blood of women after delivery (Figure 2). We sat and looked at photographs such as this and said to ourselves,

" If the foetal cells were ABO incompatible they would not be there ; they would have been got rid of by the anti-A or anti-B. Why not, therefore, try to get rid of them, since they are Rh positive, by giving anti-D ? "

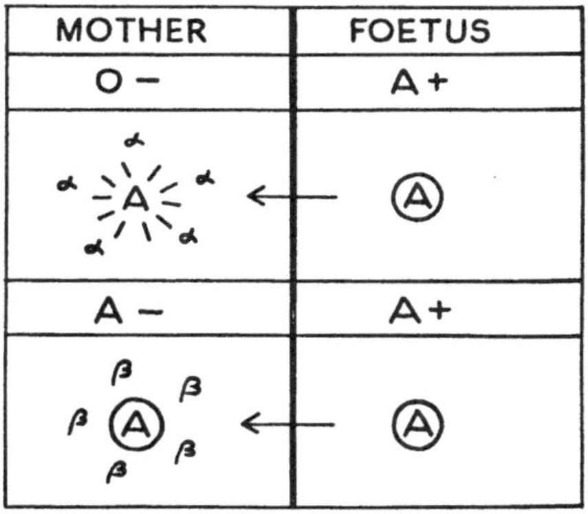

MOTHER	FOETUS
O −	A +
α ∖ι∕ α α ⟶ A ⟵ α α ∕ι∖ α ←	Ⓐ
A −	A +
β β β Ⓐ β β ←	Ⓐ

FIGURE 1

Maternal-foetal interactions in relation to the ABO and Rh blood groups of mother and foetus.
Method by which the ABO incompatible, Rh positive foetus is prevented from immunizing the Rh negative mother.
Maternal-foetal incompatibility on the 11ABO system.
Maternal-foetal compatibility on the ABO system.

Our original experiments were carried out in two directions :

1. To find out whether by giving them anti-D we could prevent Rh negative male blood donor volunteers from becoming sensitized, when Rh positive red cells were injected into them.

2. To find out at what stage of pregnancy effective transplacental ' bleeds ' occurred.

After a good deal of experimentation we found that we could usually protect the male volunteers, and that the majority of effective bleeds occurred either just before or during delivery, and that the bigger the bleed the more likely was the Rh negative woman to produce immune Rh antibodies

(Table I)—though there was, and still is, a good deal of controversy about this matter (see Cohen and Zuelzer [2]). About

TABLE I

RELATION OF FOETAL CELL COUNT AFTER DELIVERY TO SUBSEQUENT
RH IMMUNIZATION AS MEASURED SIX MONTHS AFTER DELIVERY

Cell count	0	1–4	5–60	60+	Total
Antibody	5	11	9	3	28
No antibody	188	121	44	3	356
Total	193	132	53	6	384
% with antibody	2·59	8·33	16·98	50	7·87

two years ago, however, we thought that our results were good
enough to warrant a clinical trial of anti-D in Rh negative
women. What we did was to take alternate primiparae and
give them, within forty-eight hours of delivery, 5 ml of anti-D
gammaglobulin intramuscularly, and the alternate cases were
kept as controls. Anti-D gammaglobulin (as opposed to
anti-D serum) had been suggested by the USA workers [4] in
1962 and was a considerable advance on serum which we had
used previously. The idea of giving the gammaglobulin was
to destroy any Rh positive cells which might be circulating in
the Rh negative mother after the birth, and which might, in
the normal way, immunize the mother and affect subsequent
babies. The reason we restricted cases to primiparae was to
eliminate difficulties over sensibilization (that is, sensitization
without the presence of overt immune antibodies) which could
have resulted from a previous pregnancy. Furthermore, in
the trial we included only patients whom we considered to be
' high risk ' cases, namely those who had a foetal cell score
shortly after delivery of five or more (i.e. five or more cells in a
scan of 50 low-power fields). A case such as that shown in
Figure 2, where there has been a very large foetal bleed,
would obviously be included.

Table II gives the results of what I have called the ' Liverpool Group ' trial [1] (because the centres quoted in the Table
have all followed our protocol), all the cases having been
followed up for at least six months after delivery. It will be
seen that there are differences between the various centres,

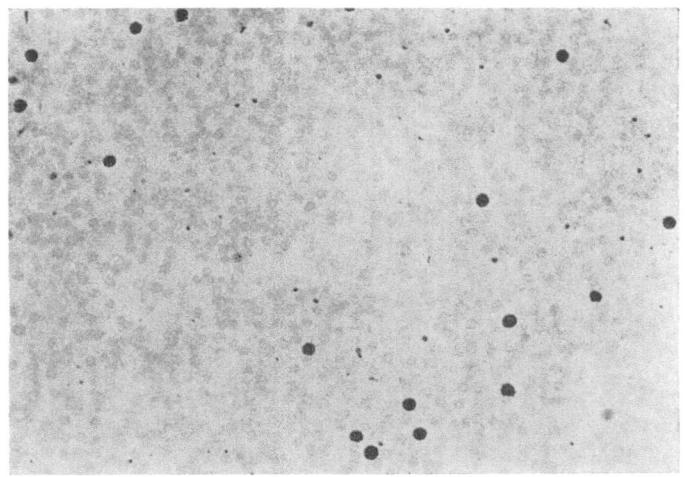

FIGURE 2
Foetal cells in the maternal circulation as demonstrated by the Kleihauer technique.

TABLE II

RESULTS OF ' LIVERPOOL GROUP ' CLINICAL TRIAL

	CONTROL CASES			TREATED CASES		
Centre	No. Immunized	Not immunized		No. Immunized	Not immunized	Doubtful
Liverpool	40	10*	30	40	0	40	0
Sheffield	15	5	10	14	0	12	2†
Leeds	5	0	5	6	0	5	1‡
Bradford	4	0	4	5	0	5	0
Baltimore	14	4	10	13	0	13	0
Total	78	19	59	78	0	75	3

* One of these had no antibody at delivery, but six months later had anti-D, detectable only by papain technique (titre 1/4). Re-tested 18 months after delivery: findings similar.

† Three months after delivery, reaction positive by papain technique, demonstrable only in neat serum. Negative Coombs test; similar findings, though weaker, at six and eight months.

‡ Tests negative at three months after delivery. Presence of anti-D queried at ten months, but all tests negative at 14 months after delivery.

Comparison of the combined control and treated groups for antibody production gives $p = 5.48 \times 10^{-7}$ if the three doubtful cases are regarded as negative and $p = 1.73 \times 10^{-4}$ if they are regarded as immunized (one-tailed tests).

but they are not statistically significant. It will at once be noticed how high is the incidence of immune antibodies in the controls; this is because we have deliberately selected ' high risk ' cases, and it is clear that estimating the number of foetal cells in the maternal blood after delivery *does* give a clue to the risk of subsequent sensitization. It is almost certain that the two doubtful cases shown in the Sheffield group are patients in whom the passively injected antibody which usually disappears after six months is still persisting for an abnormal length of time; but we cannot be absolutely sure of this.* The Leeds ' doubtful ' is probably an error; but even if we accept the doubtfuls as being immunized there is still a very large, and highly significant, difference between the controls and the treated.

Table III gives the results of similar trials by workers in the USA[7] and in Germany.[8] In the USA series, alternate cases were given 5 ml. of gammaglobulin, regardless of parity

* We now know that in both of these the antibody has disappeared and is therefore presumed to have been ' passive '.

C

TABLE III

RESULTS FROM OTHER CENTRES

| Centre | CONTROLS | | | TREATED | | |
	No.	Immune antibody present	Immune antibody absent	No.	Immune antibody present	Immune antibody absent
New York and California (Freda *et al.*, 1966, Pollack, 1966, personal communication, and Robertson, 1966, personal communication).	133	17	116	126	0	126
Freiburg (Schneider and Preisler, 1965, and Preisler and Schneider, 1966).	47	2	45	55	0	55

and also of the foetal cell score. In the Freiburg series a Kleihauer count was carried out, and all cases with a foetal cell score of one or more were used. In both these trials multiparae as well as primiparae were included; and in the German series, but not in the American, ABO incompatible pregnancies were included.

The reason for the lower incidence of antibody production in the USA controls is that the series includes both high and low risk cases. The low frequency of immunization in the German controls is partly due to the fact that ABO incompatible cases were included, and it may be that more efficient obstetrical techniques have also played a part. There is good evidence that some obstetrical procedures, particularly manual removal of the placenta, predispose to transplacental haemorrhage.

From the results it is obvious that we are protecting women up to six months after delivery, but it has been suggested by some that all we are doing is to mask antibody production. " You wait until the second pregnancies ", they say, " and you will find your treated women start producing antibodies." Well, we are still waiting. In the various series I know of, there are forty women who have been delivered of Rh positive, ABO compatible second babies and in none of those treated have immune antibodies appeared. Table IV gives the details.

TABLE IV *

ANTIBODY FORMATION IN MOTHERS INCLUDED IN THE TRIALS WHO
HAVE HAD RH POSITIVE SECOND BABIES

	CONTROLS Immune antibodies:		TREATED Immune antibodies:	
	present	absent	present	absent
UK	2	4	0	7
Germany	3	3	0	9
USA	0	6	0	6

Note. No heterogeneity present between centres.
$p = 0.011$
* See Note on page 28.

Furthermore, in the USA, Freda and his colleagues [3] have ' mimicked ' a second pregnancy in men. They injected 14 Rh negative volunteers on three occasions with Rh positive blood, the first two stimuli being followed by injections of anti-D gammaglobulin and the third one, in which a much smaller volume of blood was given ten months later, not being followed by gammaglobulin. In no case was this third injection followed by anti-D production.

There is, therefore, very strong evidence to date against the view that treatment with anti-D merely suppresses the appearance of immune antibody.

Possible dangers of giving gammaglobulin

In contrast to giving serum, there is no risk of jaundice and reactions have been almost negligible with only occasional trivial local swellings. Giving it in error to an Rh positive woman has not produced any reaction, but in the only instance we know where this was done the particular D antigen was a weak one (D^u). We still think it possible that women might become sensitized to gammaglobulin if it were necessary to repeat the injection in a number of different pregnancies, but we feel that this is a problem for the future.

Future steps

We have evidence that 5 ml. of gammaglobulin is an excessive dose for most patients and we have started a new clinical trial giving ' low risk ' cases only 1 ml. of gammaglobulin, alternate cases being kept as controls. Other centres are doing the same, and if the results are satisfactory we feel the prophylaxis should then be made generally available. Supplies of gammaglobulin have been thought to be a problem but this can be circumvented by using the technique of plasmapheresis, and it seems quite likely that it may also be possible to use serum from women who have been naturally sensitized.

An interesting point about the present clinical trial is that the results have been far better than we expected. We had thought we might protect about three-quarters of cases but we felt we should probably be unsuccessful where the effective bleed had occurred well before delivery. One reason for the unexpected result may be that, until the last few weeks of pregnancy, transplacental haemorrhages large enough to cause immunization are considerably rarer than we had thought; and we know that Rh antibodies sometimes take as long as six months or more to develop and that it is very unusual for a first Rh positive baby to be affected. A second reason for the success of the trial may be that, in contrast to the increase in titre of pre-existing antibodies which may occur in pregnancy, the primary immune response in the pregnant woman may be temporarily depressed, as it is in some animals; if this is so, red cells crossing the placenta will often not evoke an antibody response until after delivery. If such suppression of antibody response really does take place, the exact timing of a transplacental haemorrhage would be of academic importance only.

The above considerations have a bearing on the interesting work of Zipursky and his colleagues [10] who, because they thought it likely that about 25 per cent of patients become immunized during pregnancy, set out to protect this group by giving anti-D to the mother during pregnancy, a thing we had not dared to do. An initial dose of 1 ml. of anti-D gammaglobulin was given in the last trimester, followed at three-week intervals by 0·4 ml. This treatment was carried out on forty-five women at the time of the report, and thirty of them had produced healthy Rh positive infants, none of whom was anaemic, though two had a weak positive Coombs test on the red cells. We have as yet no information on how far the treatment prevents immunization, but it appears to be a safe procedure. However, it would not often be necessary if the initiation of Rh immunization is usually suppressed until after delivery.

Finally, how does giving anti-D work? There are transatlantic differences of opinion about this. We think it blocks the antigen sites on the red cells, whereas the New York group think that the gammaglobulin has an inhibitory effect on the antibody-producing cells. There are arguments for and against both points of view, but I have not got time to go into them here. Perhaps the mechanisms are not mutually exclusive. If antibody-producing cells are inhibited, the principle might have applications in the prevention of the rejection of tissue transplants in Man and also in the prevention of some types of auto-immune disease. This is speculation, however, and all I will say at present is that the evidence seems strongly to favour the view that Rh haemolytic disease will soon be conquered; and it is pleasant to feel that we can occasionally outwit our inheritance.

REFERENCES

1. CLARKE, C. A. *et al.* 1966. Prevention of Rh-haemolytic disease: results of the clinical trial. A combined study from centres in England and Baltimore. *Brit. med. J.* ii, 907.
2. COHEN, F. and ZUELZER, W. W. 1964. Identification of blood group antigens by immunofluorescence, and its application to the detection of the transplacental passage of erythrocytes in mother and child. *Vox Sang. (Basel)* 9, 75.
3. FREDA, V. J., GORMAN, J. G. and POLLACK, W. 1966. Rh factor: prevention of isoimmunization and clinical trial on mothers. *Science* 151, 828.
4. GORMAN, J. G., FREDA, V. J. and POLLACK, W. 1963. Intramuscular injection of a new experimental gamma$_2$ globulin preparation containing high levels of anti-Rh antibody, as a means of preventing sensitization to Rh. In *Proc. Cong. Int. Soc. Haematol., 9th Sept. 1962.* New York, Grune and Stratton.
5. KLEIHAUER, E., BRAUN, H. and BETKE, K. 1957. Demonstration von fetalem Hämoglobin in den Erythrocyten eines Blutausstrich. *Klin. Wschr.* 35, 637.
6. LEVINE, P. 1943. Serological factors as possible causes in spontaneous abortions. *J. Hered.* 34, 71.

7. POLLACK, W., GORMAN, J. G., FREDA, V. J., JENNINGS, E. R., SULLIVAN, J. F. and HILL, G. H. 1966. Clinical evaluation of Rh immunoglobin in the prophylaxis of immunization to the Rh factor. *Proc. Int. Soc. Blood Transfusion.* [In Press.]
8. PREISLER, O. and SCHNEIDER, J. 1966. Die Prophylaxie der Sensibilisierung der Rhesus-negativen Frau mit Anti-D-Seren und Anti-D-Globulinen. *Bibl. gynaec.* fasc. 38., 1. S. Karger (Basel).
9. ZIPURSKY, A., HULL, A., WHITE, F. D. and ISRAELS, L. G. 1959. Foetal erythrocytes in the maternal circulation. *Lancet* i, 451.
10. ZIPURSKY, A., POLLOCK, J., CHOWN, B. and ISRAELS, L. G. 1965. Transplacental isoimmunization by foetal red blood cells. *Birth Defects Original Articles Series,* 1, 1.

NOTE ADDED IN PROOF, MARCH 1967

From all centres the control series at six months after delivery shows 59 immunized out of 451 cases compared with none immunized out of 446 treated. The figures for subsequent pregnancies (see Table IV) are now 9 further immune antibodies in 32 controls compared with none in 36 treated.

FAMILY GROWTH AND ITS EFFECT ON THE RELATIONSHIP BETWEEN OBSTETRIC FACTORS AND CHILD FUNCTIONING

RAYMOND ILLSLEY

Department of Sociology, University of Aberdeen

WE know that women from different socio-economic groups also differ in their family building behaviour—in their age at first pregnancy and in the size and spacing of their families. We also know that each combination of social class, age and parity carries with it a characteristic pattern of obstetric risk. A further set of hypotheses relates obstetric risk to the intellectual functioning of the children. I propose to examine here some of the social processes underlying these sets of relationships and to show how some supposedly biological phenomena may be interpreted sociologically.

The data arise out of a major study being conducted by the Aberdeen unit on the relationship between complications of pregnancy, labour and delivery and the later intellectual functioning of the resulting children. Those who attended the corresponding Symposium of the Eugenics Society last year will be fully aware of the considerable body of hypotheses relating pregnancy experiences with intellectual functioning, and of the research reports on which they are based. These reports vary greatly in their foci, the representativeness of their populations, and their level of specificity, and also in the range, detail and accuracy of their measurements, both of obstetric events and of child functioning. Despite some contradictory findings, the bulk of the evidence leaves no doubt that an association exists. There is now evidence suggesting an association between poor intellectual performance and a variety of obstetric and foetal precursors—low birth weight, early delivery, pre-eclampsia, ante-partum haemorrhage, foetal distress, etc. In some instances the connection is clearly one of cause and effect, where specified events occurring in pregnancy or labour can be linked to specific aspects of the child's physical and neurological condition. More frequently, however, the association is conjectural, based, for example, on the lower mean IQ scores made by groups of children resulting from disturbed pregnancies of various kinds, or on the presence among mentally subnormal populations of an excess of children of low birth weight or some other obstetric complication. In some of these cases the obstetric event may indeed have led directly to lowered functioning, but more often it has not been possible to identify in the individually damaged child or the causal mechanism, and to disentangle from a complex of potentially damaging factors direct proof of a causal connection. Our own data, for example, show that many children of low birth weight have extremely high intelligence test scores and that children resulting from operative delivery have higher mean scores than those delivered spontaneously.

The problems besetting any attempt to chart the aetiological processes intervening between a presumed damaging experience and its developmental consequences are numerous. Except in extreme instances, intellectual functioning is normally measured, and reliably measured, many years after the initial insult. Between the two points in time lies a period of environmental experience, in terms both of physical and emotional health and of cultural background, which might easily dominate or obscure minor degrees of association. Tests themselves have a cultural component which cannot be eliminated and may reflect a combination of genetic and social elements whose relative weight it is impossible or meaningless to assess. Furthermore, obstetric complications themselves do not occur at random ; their incidence varies from social class to social class, between different parity and maternal age groups and particularly between different combinations of class, age and parity. Behind each complication there lies a characteristic social distribution and a corresponding social significance. The possibility therefore arises that the admitted associations between pregnancy experience and intellectual functioning arise through the operation of a third set of factors which influences both ends of the association and that the association is indirect and not causal.

To explore these various possibilities we have linked the birth records and the educational achievement of all children born in the city of Aberdeen in the period October 1950 to September 1955 and still resident in the city in December 1962, when our inquiry began. Detailed birth records exist for all children born in the city during that period, collected for research purposes and hence showing a high degree of uniformity in observation and recording. Social data on each family were recorded simultaneously and supplemented in our follow-up inquiry in 1962–3. In December 1962 we collected data about the health and the educational achievement of all children in city schools, and we linked each child's record to its birth chart. The Local Education Authority conducts routine testing of all children aged seven, nine and eleven, using standard group tests, and for the purpose of this paper I shall use the test carried out at age seven, which correlates at the 0·8 level with tests at age nine. The test used is the Moray House Picture Test, which I would describe as a test of abstract reasoning, of ability to classify, and to perceive sequences, similarities and patterns. Residential institutions inside and outside Aberdeen were searched for Aberdeen-born children whose parents were still resident in the city, and thus comparable information was obtained.

Out of the 13,800 children born between October 1950 and September 1955 who were living in the city in December 1962, we could not trace the birthplace of 115. Birth records and test scores were identified for 11,280 children who were born in the city. The remainder comprise: children born elsewhere; girls from three private schools who did not carry out the Education Authority's testing programme, and children of illegitimate birth who are being studied separately.

Detailed clinical examination of each pregnancy complication is being undertaken. In this paper I wish to focus attention on the relationship of test score to three socio-demographic factors. The first is social class, measured strictly according to the Registrar-General's 1951 Classification of Occupations. The second is the number of pregnancy which can be considered

and interpreted from two viewpoints : on the one hand we may
focus on its obstetric significance, bearing in mind the particular risks attached to first pregnancies (e.g. pre-eclampsia)
and to the higher orders of pregnancies when again a number
of characteristic complications, such as APH, begin to appear ;

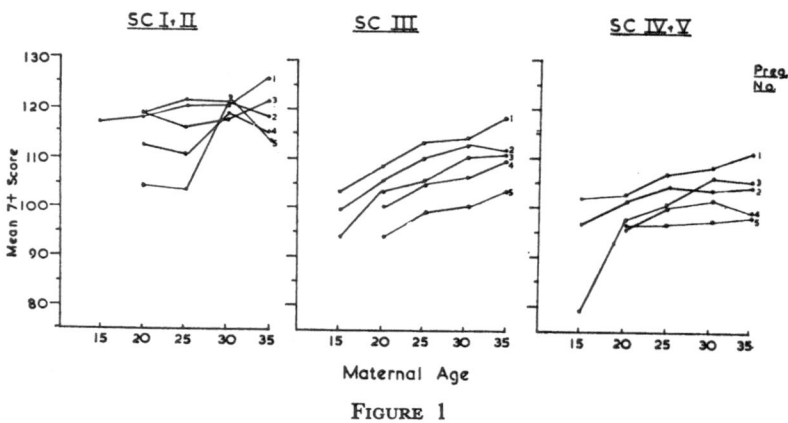

FIGURE 1

Relationship of test score to social class, maternal age and pregnancy
number

alternatively we may focus on its sociological meaning, remem-
bering that the number of pregnancies to any woman is heavily
influenced by social factors such as age at marriage, level of
education, level of aspirations, access to knowledge, husband-
wife relationships, etc. The third factor, maternal age, also
has a dual significance : on the one hand we may abstract its
physiological meaning and think in terms of immaturity in
very young women, or of physiological ageing and its implications for the intra-uterine environment, or difficulties of labour.
Particularly important are the combinations of age and number
of pregnancy : primigravidae of eighteen and thirty-eight, for

example, are quite different categories in both sociological and obstetric terms.

The relationship of test score to these three factors is illustrated in Figure 1 in which social class is based on occupation

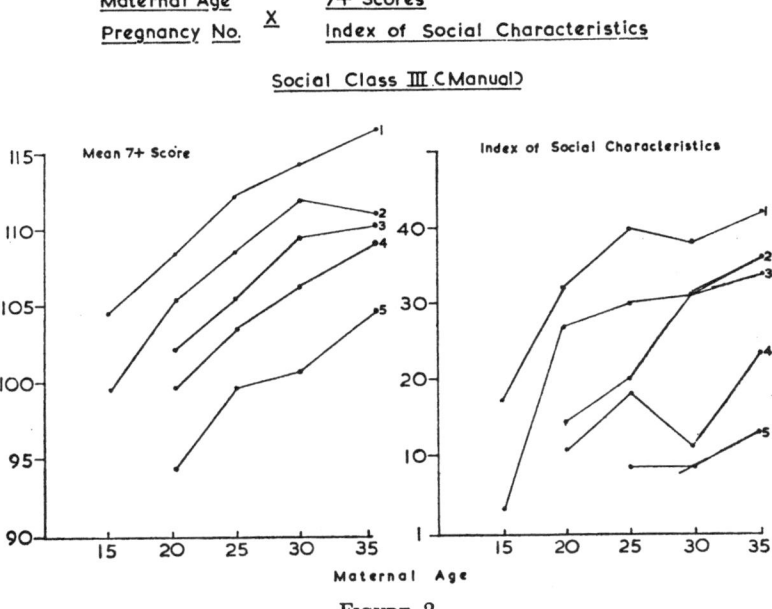

FIGURE 2

Relationship of test score to maternal age and pregnancy number
in social class III

of the husband classified according to the system used by the General Register Office.[1]

First it is clear that the general level of test scores is highest in the upper Social Classes and lowest in Social Classes IV and V, the children of semi-skilled and unskilled workers.   If we take a particular class (and for purposes of exposition I take the largest group, Social Class III) we see that at each maternal age scores are highest for children of first pregnancies and fall with increasing number of the pregnancy.   Within a given number of pregnancy, scores rise with maternal age.   These findings are less marked for Social Classes I and II and for Classes IV and V, although the broad trends are still clearly discernible.

To explore this relationship more fully, I present in Figure 2 the same data for a single socio-economic group, skilled manual workers. On this occasion I wish to consider the combined influences of age and number of the pregnancy. If we look at the age group 20–24 we see that the children of first pregnancies have a relatively high score, one which is just above 107·8, the city mean for this test. Children resulting from second pregnancies to mothers of this age have a lower mean score, and indeed the mean falls with successive pregnancies, so that children of fifth or subsequent pregnancies have a very low mean score of 94. The highest mean score occurs in children of first pregnancies to women of thirty-five or more and again falls with increasing pregnancy number within that maternal age group.

There are a number of possible explanations for this pattern of test scores. One explanation is suggested by the right hand side of the diagram. This is a crude index of social characteristics based on two criteria: the woman's own occupation, and the status of the residential area of the city in which the family were living in 1962, the date when test results were collected. The highest index score would be obtained by a professionally qualified woman living in the West End of the city, the lowest by an unskilled worker living in an inter-war tenement estate occupied by the ex-residents of a cleared slum area. There is a general parallelism between the two diagrams. Again in the 20–24 age group, women having their first babies at that age have a high index score, and the score falls with increasing pregnancy number. In this socio-economic group first pregnancy occurs more frequently in this age range than in any other five-year period; fourth or subsequent pregnancies on the other hand are comparatively rare, and, as the index of social characteristics suggests, they occur to a selected socio-economic group. One interpretation of the data, therefore, is that variations in test-scores between maternal age and number-of-pregnancy groups merely reflect hidden socio-economic influences—the selective factors which determine the social composition of groups which, for example, have one child by age thirty-five compared with those who have four by age twenty-five—education, aspirations, income, ability to manipulate their own environment through family planning, and so on.

At this point we may ask whether the general pattern of

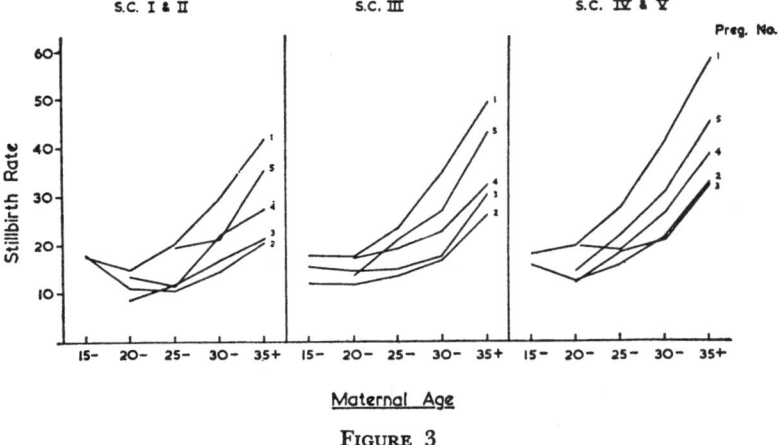

Social Class
Maternal Age X Stillbirth Rate
Pregnancy No.

FIGURE 3

Relationship of stillbirth rates to social class, maternal age and pregnancy number

test scores is indicative of a strong physiological or obstetric influence. Perhaps the best indicator of physiological risk is provided by the incidence of stillbirth rates in the various combinations of social class, maternal age and pregnancy number. These are shown in Figure 3 adapted from the data for England and Wales at the 1951 Decennial Census and published in the report of Heady and Heasman.[2] They differ from my data in two respects—first, they refer to England and Wales, and secondly, they show parity as opposed to number of pregnancy. The pattern shown in this diagram is, however, closely parallel to the Aberdeen stillbirth rates of that period and I have chosen to present them simply because with larger numbers the patterns emerge more clearly.

I wish to make three main points concerning these rates.

1. They show one result in common with the test scores reported earlier. Stillbirth rates are high in the lower social class and fall with increasing socio-economic status. In other words, at each age and parity level the chances of survival are associated with the socio-economic condition of the population. At one end of the scale we have high stillbirth rates and low test scores, and at the other end low stillbirth rates and high test scores.

2. A major discrepancy occurs, however, when we turn to the number of pregnancy or parity. At all ages and in all classes stillbirth rates are highest for first pregnancies. They fall to a low point in second and third pregnancies and rise steadily thereafter. The high stillbirth rate in first pregnancies is in direct opposition to their relative socio-economic status. On a whole variety of indicators—class of upbringing, the wife's own occupation, educational level and maternal stature— first pregnancies rank higher than any other. They alone represent a true cross-section of childbearing women, subsequent pregnancies becoming increasingly socially biased, as women who have only one, two or three pregnancies cease to be represented. Such social bias is only partially reduced by considering each social class separately because, with existing crude methods of classification, each class contains women with a very wide range of social characteristics. This point emerged in Figure 2 (showing the distribution of scores on an index of social characteristics), where within the group of skilled manual workers, first pregnancies ranked highest at each age and parity. Physiological risks are therefore high in first pregnancies despite their socio-economic characteristics. The pattern of test scores, on the other hand, betrayed no evidence of increased risk to the intellectual functioning of the child, the risk being higher in first than in any other pregnancy.

3. Turning to maternal age, we again see a distinctive pattern of risk within each social class and parity. From age 20–24 onwards there is a marked and consistent rise in the

stillbirth rates with increasing age. Rates at age thirty-five and over are frequently three, four or five times as high as at 20–24. Rates for the youngest age group, on the other hand, show a tendency to be slightly higher than those for the 20–24 age group. Increase of liability to foetal loss with increasing age undoubtedly results very largely from physiological ageing, from a lowering of reproductive efficiency and a reduced

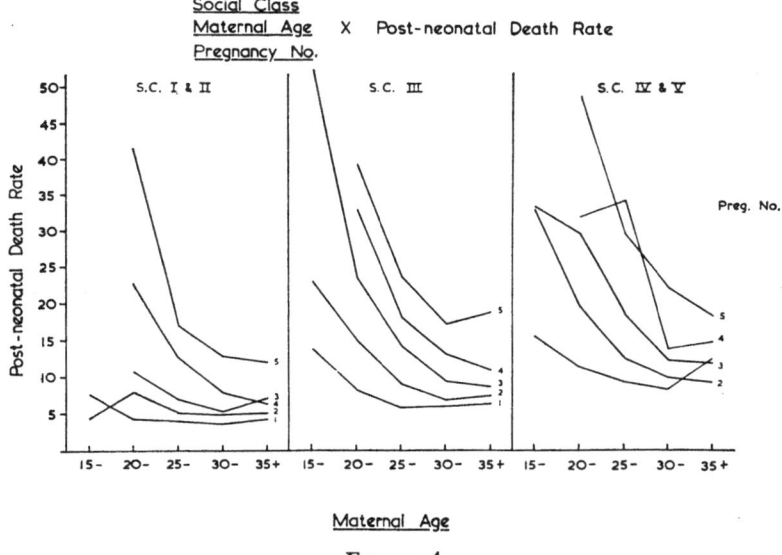

FIGURE 4

Relationship of post-neonatal death rate to social class, maternal age and pregnancy number

capacity to absorb the strain of pregnancy and its complications without transmitting ill-effects to the developing foetus. This again is in direct opposition to socio-economic status, and to the test scores of the surviving child.

As a model of physiological or obstetric risk I have taken stillbirth rates. For a model of socio-economic risk I now turn to post-neonatal death, which has long been regarded as a most sensitive indicator of poverty and its concomitants. Data, drawn from the same source, again refer to England and Wales (Figure 4).

Again three points may be stated, this time very simply and briefly :

1. Rates from Classes IV and V are highest, and fall sharply with increasing socio-economic status.

2. Rates rise consistently with increasing parity.

3. Rates fall consistently with increasing age.

This, rather than the physiological model, conforms with the pattern of test scores.

What I have shown in these few Figures does not disprove the hypothesized causal relationship between pregnancy experience and intellectual functioning. It is quite possible that within these broad groupings and patterns there are a number of mentally subnormal children whose condition is directly attributable to obstetric factors. It is also possible that in other areas, where obstetric practice is less sophisticated, damage may occur frequently. To explore such problems requires a different and more detailed type of analysis ; this we are endeavouring to achieve in other parts of our research programme.

We can, however, draw certain conclusions.

1. Whatever association exists in our type of society is likely to be small, affecting very limited numbers of individuals, rather than large population categories, and it is heavily outweighed by other genetic or social influences. Nowhere are obstetric risks higher than in first pregnancies to elderly women, but the offspring who survive function intellectually at a very high level.

2. In any further analysis of this problem we cannot afford to ignore the known associations of IQ with maternal age and number of pregnancy, and the social environment of mother and child. Preliminary analyses of our data show time and again that children resulting from pregnancies disturbed in a particular way have test scores higher or lower than those of ' normal ' children. On the other hand, when we take into account the distribution of the complication in terms of social class, age and parity, the difference frequently disappears or is reduced to insignificance. This is not always the case, however ; even when such allowances are made, there are still a few conditions which warrant closer attention. But at least,

when such allowances have been made, we know that a real problem exists.

3. Finally I want to enter a cautionary note. The distribution of test scores by number of pregnancy shown earlier may leave a suggestion that first children have higher intellectual abilities than second children and that intelligence test scores decline thereafter with position in family. This can be checked by looking simultaneously at test scores by family size and ordinal position in the family (Table I). If we consider a particular family size we find no variation by ordinal position either in the whole sample or a particular social class. In families of five, for example, mean scores from the first to the fifth are 100·8, 100·4, 100·6, 100·8 and 102·0.

TABLE I

MEAN TEST SCORES BY FAMILY SIZE AND ORDINAL POSITION

FAMILY SIZE	ORDINAL POSITION				
	1st	2nd	3rd	4th	5th or later
	Mean Test Score				
1	112·5				
2	111·6	112·4			
3	108·7	109·1	109·2		
4	104·8	104·2	103·9	107·1	
5+	100·8	100·4	100·6	100·8	102·0
	Numbers				
1	880				
2	1476	1501			
3	837	886	795		
4	328	417	445	378	
5+	158	250	304	360	508

It is true, however, at all ordinal positions, that test score falls with increasing size of family. The decrease in test score by number of pregnancy shown in the figures presented earlier arises indirectly through the mediation of family size. Fourth pregnancies, for example, can occur only in mothers who have families of four or more. First pregnancies, however, will include not only mothers who may ultimately have a further

D

five, six or more, but also those who limit themselves to one, two or three.

The reasons for the test scores being higher in smaller families are complex. On the one hand, there is the possibility that the smallness of the family itself contributes to a higher educational achievement in the child either through permitting a better material environment or because parents are able to devote more attention to the child and thereby offer a more stimulating set of experiences. But the process starts further back—in the education, aspirations and planning habits of the parents. Where these are at a high level, the size and spacing of the family is more likely to be tailored to its expected resources, and to the parents' ability to raise their children to their own high standards and expectations. At the opposite extreme—with low educational levels in the parents, poor material resources and prospects, and lower ability or motivation to adjust means to ends—family size is frequently inversely correlated with income, wealth and security. In this connection, it is interesting that we found a slight fall in the test scores of children in Social Class I with increasing family size, but a marked fall in those of children in the more heterogeneous Class II and in Class III (non-manual). It is most marked of all in Classes III, IV and V. It would appear that in Social Class I children are at no disadvantage in families of four or five as compared with children in families of one, two or three. This may be because their cultural and material environment gives them adequate support, but this in turn depends on family-building habits; most Class I parents who have large families have them, not because they cannot control their reproductive habits, but because they want large families and know they have the means to support them. In Classes IV and V, on the other hand, family size is frequently not willed but accidental and a large family is, in general, indicative of poor material standards and of the parents' inability to control their own environment.

Conception, pregnancy, delivery and child development are linked and unified by the family and social context in which they occur. Items of sexual and family behaviour can be abstracted, given a physiological connotation, and called age-and-parity. Whatever we call them, they do not lose their sociological significance, and in any examination of obstetric-child functioning relationships we should remember that what appears primarily a physiological phenomenon may well have an alternative explanation.

DISCUSSION

PROFESSOR ILLSLEY, a questioner said, had described a relationship between maternal age within parity and IQ at seven. Was there a similar relationship between height and IQ at seven ? In trying to think about the very intricate relationships with IQ, it sometimes seemed easier to consider something a little less intricate, though obviously subjected to very similar sociological and physiological forces. Professor Illsley replied that a similar relationship with the child's height had been found ; it was not nearly so marked as the relationship with intelligence test scores, but it clearly did exist.

He had been interested, too, when looking at a group of families and the way in which they had gone about the whole process of family planning, to find that there was a relationship between the process of family building and a number of criteria about the child. Test score and the height of the child were both related to the planning habits of the parents.

On the question of adopted children, Professor Illsley said that he had some information and was looking at it very carefully, but there were snags here. Out of this population he had so far been able to identify only sixty children, who had been adopted into a different family and where he knew something about the social backgrounds of the natural and the adoptive parents.

For what it was worth—and there were a lot of difficulties—this showed two or three things fairly clearly. First, these adopted children, all of them originally illegitimate, had a mean IQ approximately three points higher than the city mean. The pattern of their scores tended to follow that of their adoptive parents rather than that of the parents who had given birth to them. He added that an interesting point was that illegitimate children who had been *adopted* into social Class V had scores much higher than children *born* into Class V, higher in fact than the total city mean. They had probably been adopted into very selected families. A difficulty with data of this kind was that we could not assess the effect of selection processes, carried out by the parents themselves and by adoption societies.

REFERENCES

1. GENERAL REGISTER OFFICE. 1951. *Classification of Occupations*, 1950. London. H.M.S.O.
2. HEADY, J. A. and HEASMAN, M.A. 1959. *Social and Biological Factors in Infant Mortality*. London. H.M.S.O.

SOCIAL AND BIOLOGICAL INFLUENCES ON FOETAL AND INFANT DEATHS

Thomas McKeown

Department of Social Medicine, University of Birmingham

In this paper I propose to discuss the major influences responsible for foetal and infant deaths and to assess the extent to which these deaths are potentially preventable. This undertaking poses three main difficulties. The first is that a considerable number of deaths are attributed either to ill-defined causes, or to pseudo-specific causes, such as immaturity— which only conceal our ignorance. A second difficulty arises because in some cases where death is associated with a well-recognized condition, such as toxaemia of pregnancy or a congenital malformation, little or nothing may be known about its aetiology. Finally, even where there is considerable knowledge of aetiology, as in the case of a post-natal infection, death cannot without reservation be considered invariably preventable. It is this uncertainty which has led to the use of the somewhat ambiguous term ' potentially preventable ', which at least has the excuse that it does not claim too much.

Let us consider first the change in age-distribution of deaths in England and Wales which has resulted from improvements in health during the past century (Figure 1). With due regard for uncertainty about the frequency of abortion—we can be confident only that it was high in the past and remains relatively high to-day—the main features of the trend since 1838–54 are : a moderate reduction of prenatal deaths ; a large reduction in deaths between birth and age 45 ; a small reduction at age 45–64 ; and an increase at ages 65 and over. This increase is of course due to the larger number of people surviving to late life.

Two general points should be made in relation to the age period with which we are concerned.

a. Mortality is very much higher before birth than between birth and age fifteen. Moreover the significance of pre-natal influences is even greater than this comparison suggests, since many post-natal deaths are due to conditions before or during delivery.

b. Most pre-natal deaths are attributable to the uterine

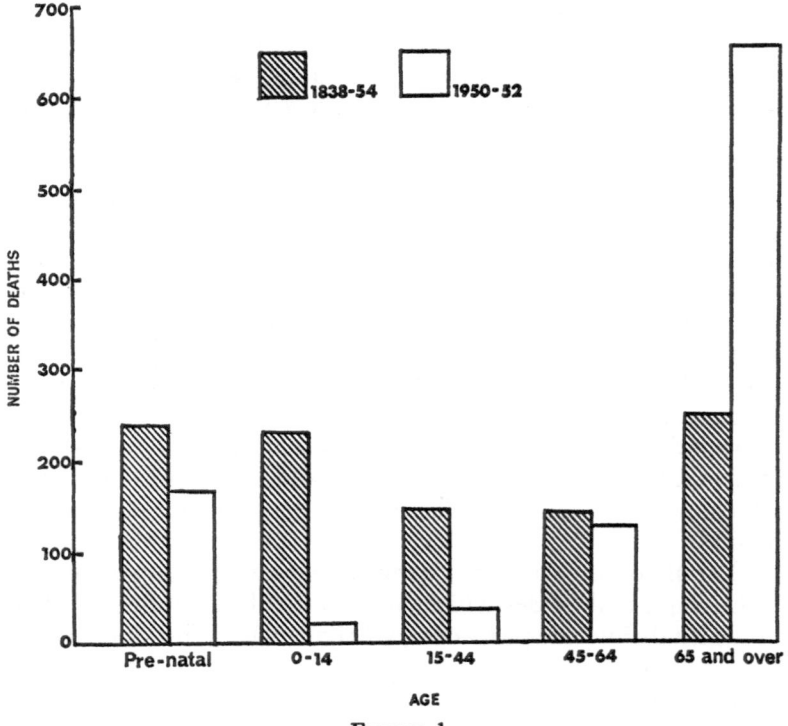

FIGURE 1

Mortality (per 1000 conceptions) at different ages.   England and Wales

environment rather than to lethal genes. Medawar [1] drew attention to the point that, since genes associated with death before the end of reproduction tend to be eliminated, deaths in this period must be largely of environmental origin. There could be no stronger support for this conclusion than the profound reduction of mortality in the period early to middle life which has resulted from changes in the environment without selection of parents. Since selection operates even more strongly before birth, there are equally good reasons for believing that most pre-natal deaths are due to the uterine

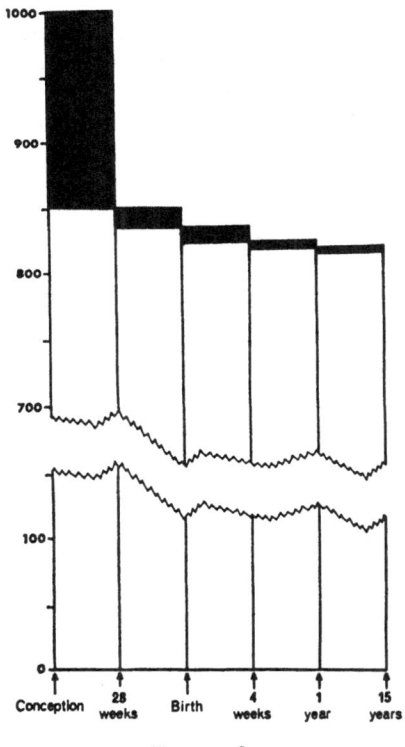

FIGURE 2

Estimated number of deaths (in black) per 1000 conceptions. England and Wales, 1964

environment. Unfortunately it is much more difficult to control than the post-natal environment.

Figure 2 shows for England and Wales, 1964, the extent of mortality in different periods between conception and age fifteen. The only uncertainty again concerns the frequency of abortion, and the estimate is justified only to make the point that the number of deaths between conception and 28 weeks (when stillbirths are registered) is undoubtedly much higher than between 28 weeks and age five. Because of variation in time intervals, the reduction of mortality with increasing age after 28 weeks is greater than the Figure indicates. Clearly the uterus is the most dangerous environment in which the individual is placed between conception and late life. Far from wishing to crawl back into it, as some psychiatrists suggest, we should consider ourselves fortunate to have got out alive.

CAUSES OF DEATH

The next four figures (3–6) attempt to put into perspective the major causes of death between 28 weeks and age fifteen in England and Wales, 1964. For this purpose it has been desirable to group causes according to the feasibility of their prevention.

Accidents and Infections

In this class are included deaths due to infections and ' accidents and violence '. Although very different aetiologically, they have in common the fact that they are largely preventable. They contribute little to stillbirths and relatively little to mortality in the first four weeks after birth (Figure 3). On the other hand, together they account for considerably more than half the mortality after the first month and before age fifteen. I shall not attempt to discuss in detail the problems presented by the common infections and accidents and violence. Enough is known to justify the conclusion that they are potentially preventable.

Obstetric conditions

Under this heading we have grouped a number of different causes of death which need some separate comment.

a. Maternal toxaemia and other maternal disease. On present knowledge, toxaemia of pregnancy cannot be regarded as being invariably preventable ; but in some cases its effects

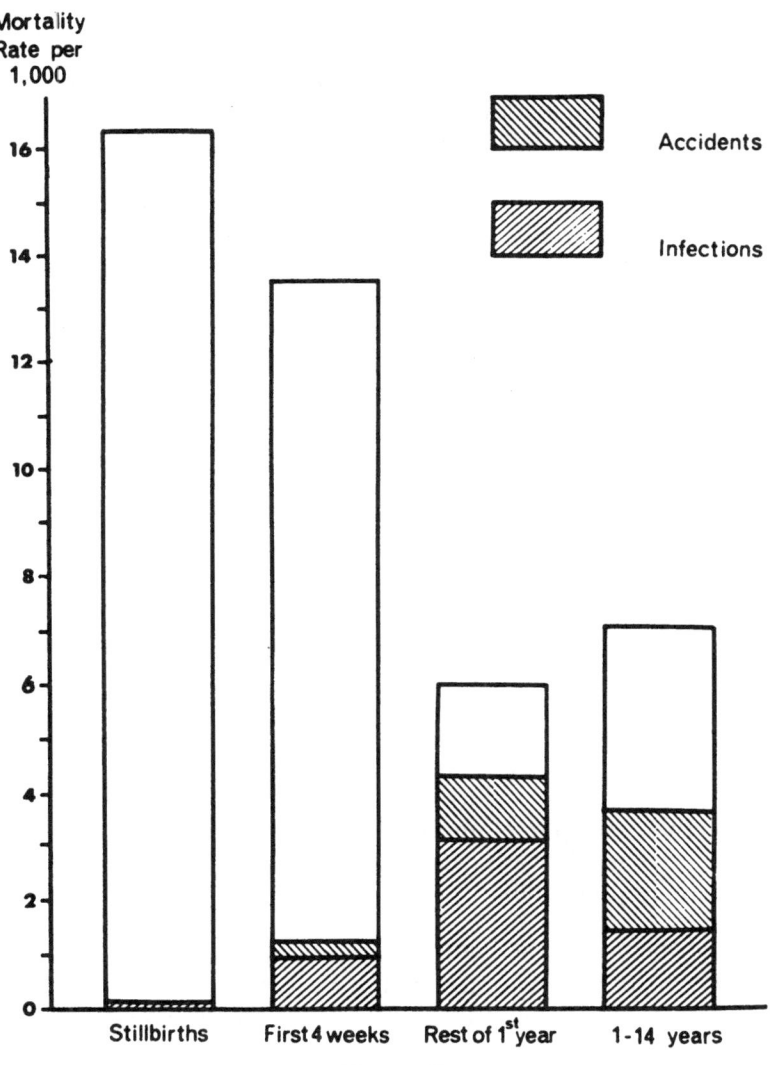

FIGURE 3

Contribution of ' accidents and infections ' to mortality before age 15.
England and Wales, 1964

on the foetus can be avoided by good ante-natal care, and if knowledge of aetiology were more complete the disease might prove to be almost wholly preventable.   The same considerations apply also to the other less frequent maternal diseases which still account for some foetal deaths.   Their incidence has been substantially reduced by improvement in women's health, and there is no reason to doubt that there is scope for further reduction.

   b. Cord and placental conditions, difficult labour, birth injury, asphyxia, atelectasis. (Asphyxia and atelectasis associated with immaturity are classified under ' immaturity '.) All these very different conditions have in common the fact that death is usually associated with delivery.   Many of the deaths are avoidable, and are in fact avoided in good obstetric practice ; investigation of the circumstances of maternal and infant deaths leaves little doubt that their occurrence is a serious criticism of the obstetric services.   Improvement will require a higher standard of obstetric care before and during delivery, and since some obstetric emergencies are unpredictable it is questionable whether it can be fully achieved without a high proportion of hospital deliveries.   In some other deaths, for example those attributed to ' placental conditions ', the circumstances of death are more obscure.

   The contribution of these ' obstetric conditions ' to mortality is shown in Figure 4.   They account for more than half the stillbirths and for a substantial proportion of deaths in the first four weeks after birth.   Thereafter their contribution is negligible.   In spite of uncertainty, it seems reasonable to conclude that most deaths attributed to obstetric conditions are potentially preventable.

Malformations and immaturity

   We have already noted that our present limited knowledge of these conditions provides little reason for thinking that they offer much scope for prevention.   The discovery of two controllable causes of malformation, rubella and thalidomide, encourages the hope that others may be identified ; and as genetic knowledge increases, a few malformations may be prevented by avoidance of pregnancy.   So far as we can judge, however, the great majority are likely to remain unpredictable. Moreover, most of the significant influences are probably within

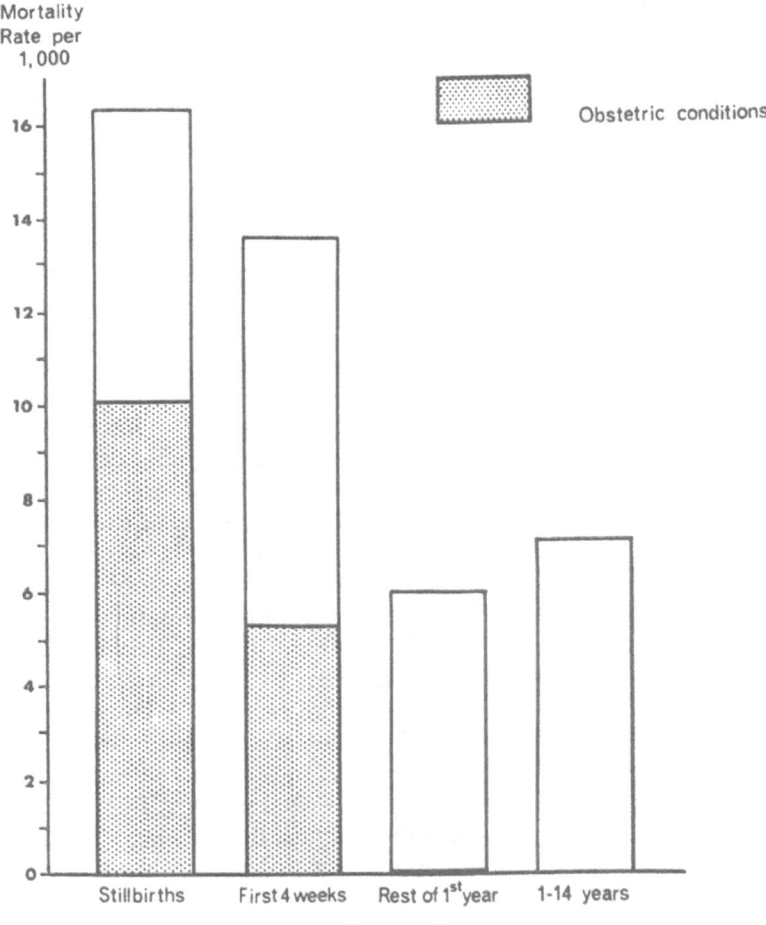

FIGURE 4

Contribution of ' obstetric conditions ' to mortality before age 15. England and Wales, 1964

the uterus rather than external to the mother, and operate so early, often before pregnancy is recognized, that their control is almost inconceivable.   Seriously malformed foetuses could sometimes be identified and removed before they are viable; but this possibility raises ethical and religious, rather than biological issues and is outside the scope of the present discussion. Nevertheless it may prove to be true that most serious malformations can neither be predicted before conception nor prevented by manipulation of the pre-natal environment, and the birth of seriously handicapped individuals, in many cases with a short expectation of life, may be avoided only by removal of the affected foetus.

The causes of immaturity are multiple and complex. In general immaturity, identified usually by low birth weight, is due to one or both of two causes: retardation of growth during pregnancy, and shortening of the period of intrauterine development by early onset of labour. The diverse problems involved may be illustrated by two examples. In multiple pregnancy foetal growth is retarded from a gestational stage at which the total foetal weight reaches about 7 lb. (Hence the time of onset of retardation varies with the number of foetuses in the uterus.) The reasons for this retardation are unknown, but it cannot be prevented by additional feeding of the mother and may be related to the blood supply of the uterus. In placenta praevia, on the other hand, foetal growth to the time of delivery appears to be about normal, and immaturity is due mainly to early delivery. The problem here is very different from that in multiple pregnancy but, so far as present knowledge goes, equally outside our control. Most cases of immaturity are not associated with any obvious immediate cause—such as multiple pregnancy or placenta praevia—and all that can be said with confidence is that many of them cannot be prevented by an improved standard of living or better obstetric services. There is of course considerable scope for prevention of death of the immature infant by improved care after birth.

Figure 5 shows the contribution of malformations and immaturity to mortality. Together they account for about 20 per cent of stillbirths, and for a substantial and, as other causes of death decrease, an increasing proportion of post-natal deaths, particularly in the first four weeks. Although little is known about the aetiology of malformations and immaturity, there have been some improvements in treatment, and it is questionable whether a worthwhile estimate can be made of the proportion of deaths which are potentially preventable.

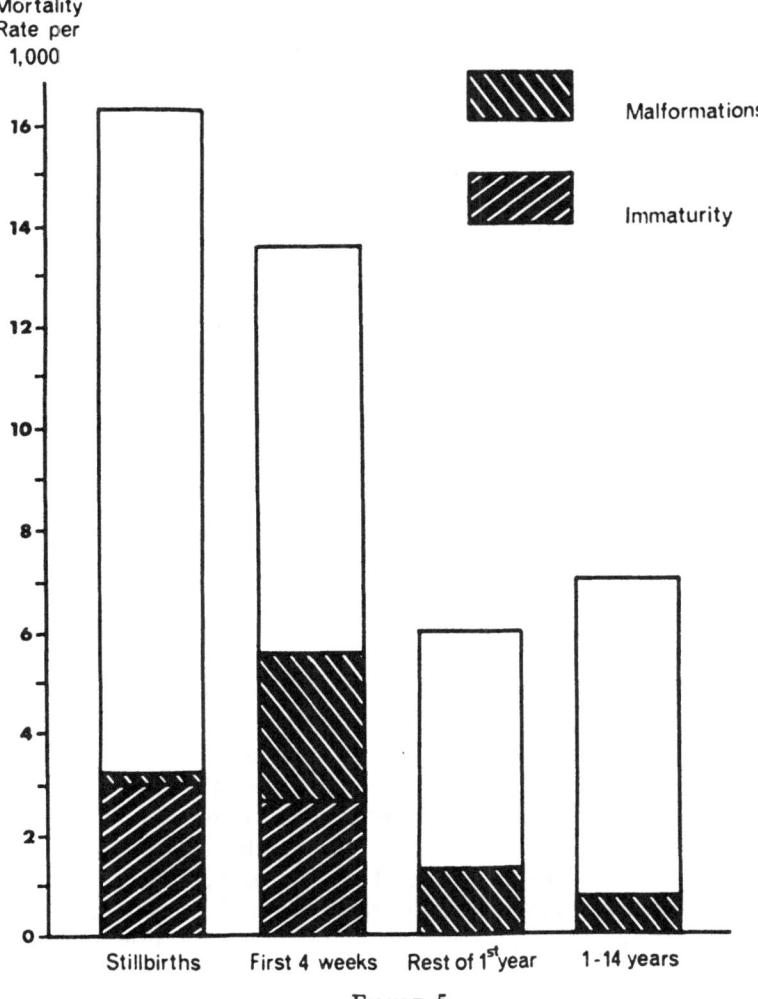

FIGURE 5

Contribution of ' malformation and immaturity ' to mortality before age 15.   England and Wales, 1964

The incidence of malformations appears to have remained at about the same level during the period in which it has been examined.

Other conditions

' Other conditions ' are an even more anomalous collection of causes of death, which are taken together because there is little evidence on which to assess their preventability. They include neoplasms, some other specified causes, and ill-defined or unspecified causes. Many of these conditions are determined before birth and are probably, in some cases certainly, not associated with the most tractable causes of foetal death: maternal disease and difficult labour. For this reason, and in recognition of our ignorance, they are considered to be deaths in which the scope for prevention is, to say the least, uncertain. Figure 6 shows their contribution to foetal and post-natal deaths.

BIOLOGICAL AND SOCIAL INFLUENCES ON MORTALITY

We must now try to summarize our conclusions which emerge from consideration of different causes of death. Figure 7 shows the contribution of each of the four classes of death (discussed above) to stillbirths and mortality between birth and age fifteen.

Although we have not been concerned with deaths before 28 weeks of pregnancy, it may be noted that a substantial proportion of deaths before that time are not of biological origin, but are due to the social, psychological and economic circumstances which lead to induced abortion. Knowledge of spontaneous abortions is very incomplete. A considerable number are malformed, some of them grossly, and completion of pregnancy would be undesirable even if it were possible. There are no means of assessing whether such conditions can be prevented, but it is a reasonable guess that most of them cannot.

In the light of our earlier comments we shall regard two of the four classes of deaths (' infections and accidents ' and ' obstetric conditions ') as potentially preventable. On this basis we should consider as preventable approximately two-thirds of stillbirths, half the deaths in the first four weeks,

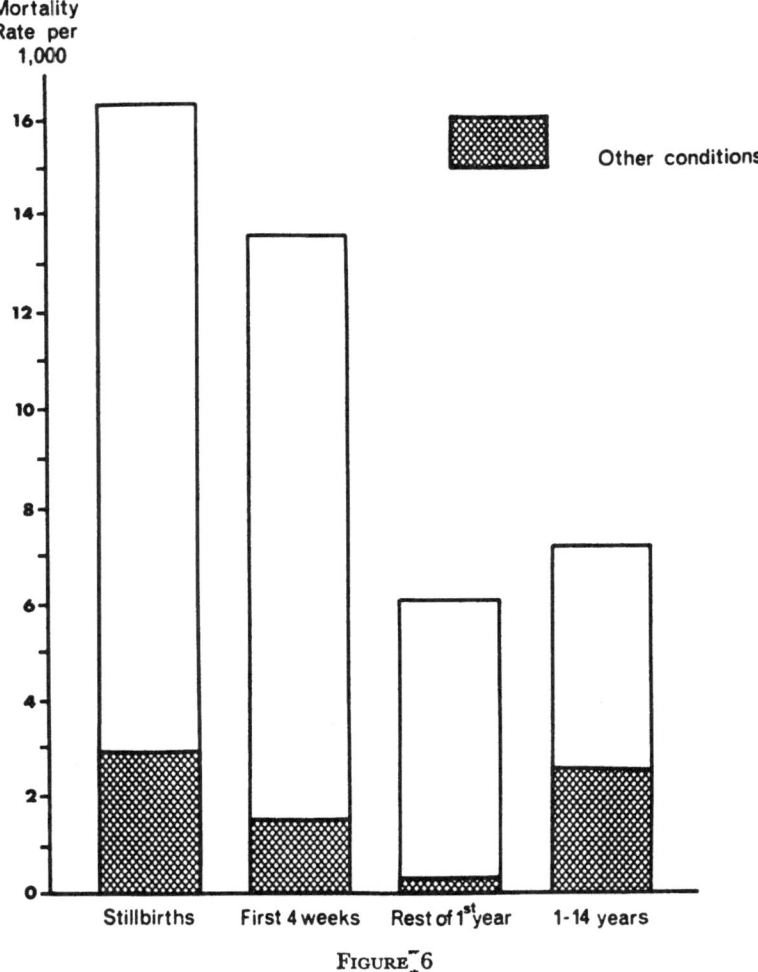

FIGURE 6

Contribution of ' other conditions ' to mortality before age 15.
England and Wales, 1964

more than two-thirds of those in the remainder of the first year, and half of those occurring from age one to age fifteen (Figure

7). In Figure 8 all deaths between 28 weeks of pregnancy and age fifteen are considered together; more than half are in the classes which offer substantial scope for prevention.

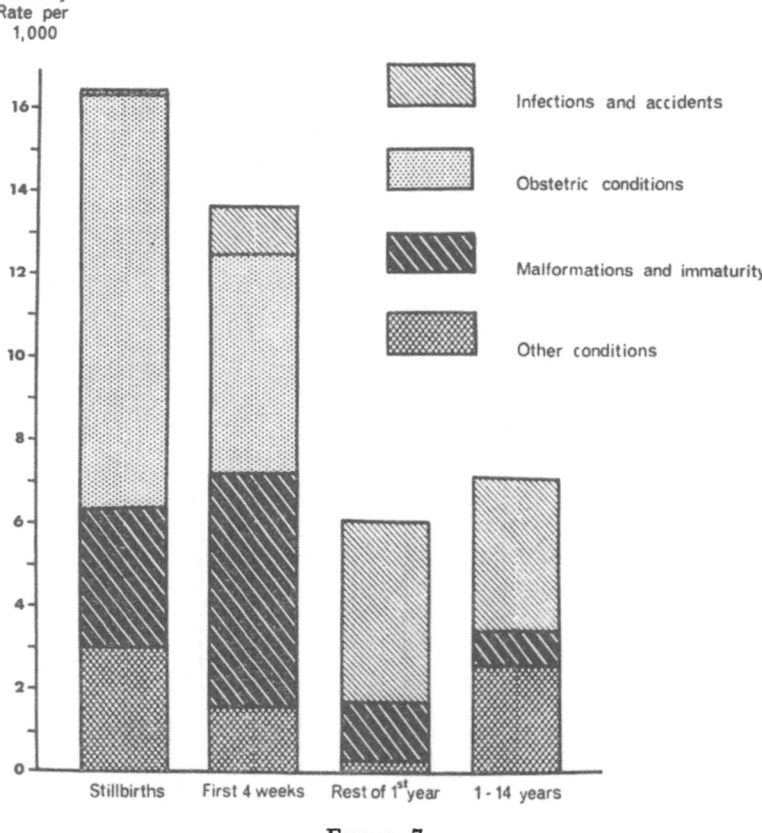

FIGURE 7

Contribution of the four classes of causes of death shown in Figures 3–6 to mortality before age 15. England and Wales, 1964

Upon what kinds of measures do these advances mainly depend ? In the case of stillbirths they require better standards of maternal health—by no means an unrealistic goal, since the standard of living is still rising—and improved obstetric care in pregnancy and labour. The latter requires both more competent obstetric management and (since special facilities as well as skill are needed for unforeseen emergencies) delivery in hospital. Obstetric care is also the most significant influence in relation to early post-natal deaths (Figure 7).

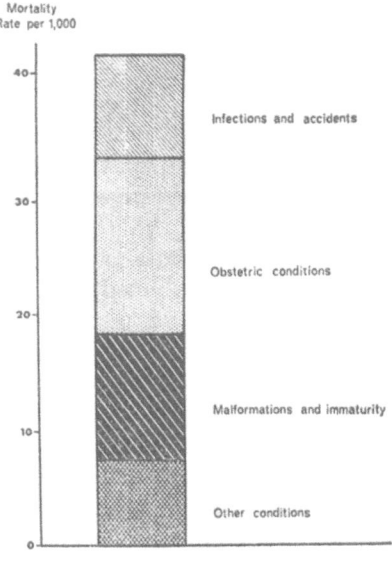

FIGURE 8

Mortality (28 weeks' gestation to age 15) according to cause. England and Wales, 1964

The measures needed to reduce mortality between one month and age fifteen are quite different. Most of those required to prevent accidents and violence are well recognized, if not always easy to apply. Infections are still the most important cause of post-natal mortality in childhood. In some diseases the measures needed to reduce it are well recognized; and the fact that in the most favourable circumstances (for example among first-born children of well-to-do parents) mortality is much lower than the prevailing general population rates, indicates the scope for further preventive advance.

E

The problems presented by the other large classes of death have not been discussed in detail. There are no grounds for thinking that with further knowledge most of them will prove to be preventable; indeed there is some reason to believe that they may not. At the present time these problems largely concern the research worker and those responsible for the services needed by the congenitally handicapped.

SUMMARY

If we may accept a very rough estimate of the frequency of abortion (150 per 1000 conceptions), there are approximately 200 deaths per 1000 conceptions before age fifteen, of which 150 are abortions, twenty stillbirths and thirty post-natal deaths (of which half occur within a month of birth). A substantial proportion of the abortions are induced, and the problems are therefore the social, economic and psychological reasons for unwanted pregnancies. Knowledge of the aetiology of spontaneous abortions is very incomplete; some of them are malformed, and continuation of pregnancy would be undesirable.

Four-fifths of the fifty deaths between the twenty-eighth week of gestation and age fourteen are due to causes operating before birth. Perhaps half of them are potentially preventable —by prevention of maternal illness (especially toxaemia); by improved obstetric care, particularly during labour; or by prevention of infections, accident and violence after birth. The remaining stillbirths or child deaths are due to causes such as congenital malformations, immaturity and neoplasm, in which the little that is known about aetiology is enough to make us cautious about the feasibility of control of the pre-natal environmental influences which are commonly involved.

REFERENCES

1. MEDAWAR, P. B. 1957. *The Uniqueness of the Individual*. London. Methuen.

SOME MAJOR CAUSES OF ILLNESS:
I. SOMATIC ILLNESS

Chairman: LORD PLATT

RECENT TRENDS IN MORTALITY

C. C. SPICER

General Register Office, Somerset House, London

THIS account of the major trends in mortality in the last ten to twenty years is necessarily limited to the broad features of these trends, owing to lack of time and space; a full treatment would be much more extensive. A good recent discussion has been published by Campbell[1] and in preparing this note I have drawn heavily on his account.

All the statistics I shall use are drawn from the various publications of the Registrar General and are based on the entries on death certificates as to cause of death. The deficiencies of such statistics are well known and must be accepted and borne in mind when considering the findings. Doctors' entries are made with widely varying degrees of accuracy and conscientiousness and are subject to changes of fashion in diagnosis and periodic revisions of methods of disease classification. In addition, to allow manageable tables to be published, the often complicated death entries have to be reduced to a single diagnosis, hopefully referred to as the ' underlying cause '. In recent years this practice has become increasingly unrealistic.

In spite of these limitations, I think it would be true to say that death certificates form the only continuous and universal source of information on mortality trends; and experience has shown that they can be a useful, indeed essential, guide to the broad trends in public health. In the last twenty years they have probably become less useful as a guide for many purposes, as advances in treatment have widened the gap between mortality and morbidity. In the next twenty years the gap will grow even wider and the importance of morbidity statistics will greatly increase, but we are still only in the very early stages of knowing how best to collect and interpret such statistics.

It is probably most convenient to go briefly through the

main ages of man and mention the major trends in each and then try to summarize the most important trends and features that are revealed. The general picture of mortality in the last twenty to thirty years has been one of decline in the overall rate, much more pronounced in the female, and in many of the individual causes of death. However, the decline in the death rate has been steadily becoming less rapid in the last two decades; and we are now in a situation where radical changes in our approach to disease, and in our knowledge of aetiology, are required to speed up again the improvement in mortality.

Infant mortality (0–1 year)

There has been a steady decline in infant mortality, i.e. mortality under one year of age, but this has not been evenly distributed over the first year. The major improvement has occurred in the period from one month to one year, and a comparatively small improvement in the first month, and still less in the first week. In recent years the post-neonatal mortality has declined much less rapidly and has even shown a tendency to rise slightly. The chief cause of the decline in death rate at all ages has been the very great reduction in mortality from communicable diseases, particularly gastro-enteritis, and respiratory disease, mainly broncho-pneumonia. Little change has occurred in the number of deaths associated with congenital malformations or immaturity complicated by respiratory disease. It is also very significant that, in spite of the great general reduction in mortality, the gap between the death rates of the different social classes still remains and has even widened, and there are still very wide variations between different areas in this country. The low death rates in other countries also demonstrate how much progress can still be made at this age group.

Pre-school age (1–4 years)

This age group has benefited more than any other from the improvements in public health during the last twenty to thirty years, but since 1955 there has been a slowing down in the rate of improvement. A great part of the improvement can be ascribed to preventive and therapeutic measures in the infectious diseases, notably diphtheria, measles, whooping cough and tuberculosis. Pneumonia is still a notable cause of death, others being congenital malformations, road accidents, and leukaemia and other neoplasms. Leukaemia and the cancers, and congenital heart disease have increased, the former probably because of improved diagnosis as well as because of the removal of competing risks, the latter probably because better medical care now prolongs survival into this age group. It is interesting that the number of deaths from road accidents at this age has been erratically declining.

School age (5–14 years)

The lowest death rates at any age occur in this group. The major changes that have taken place have been due to the eradication of infectious disease and rheumatic fever. Deaths from malignant disease have been increasing slowly and deaths from congenital heart disease rapidly. Deaths from road accidents though a major source have not been increasing, and twice as many boys are killed in this way as girls. The general death rate has remained fairly steady.

Adolescence and after (15–24 years)

The outstanding changes affecting death rate in this group have been the disappearance of tuberculosis and the rise in road accidents. The latter now account for about 40 per cent of all deaths among males at this age, and deaths from this cause are about five times as numerous among males as females. A very large part of them are associated with riding motor bicycles. The suicide rate among males has wavered erratically; it is always higher than among females, but among them it has been rising for some years. The mortality rate as a whole has been more or less stationary among males for some time, and slowly decreasing among females of this age group.

Young adult (25–34 years)

The divergence in the trends of male and female mortality begins to become very marked in this age group. Road accidents and suicide are both prime causes of death and have been rising for both sexes, though the rise in road accident deaths is not very great. The rise in suicides has been marked in the ' affluent ' period beginning in the 1950s.

Adult (35–44 years)

At this age some of the most frequent causes of death start to become manifest. In males coronary heart disease, lung cancer and suicide are prominent, and in females breast cancer and suicide, and carcinoma of the cervix also begins to show itself. It is a curious fact that lung cancer among males in this age group is now not increasing. In general death rates are low, being more or less stationary among males and declining slightly in females.'

Middle life (45–64 years)

The commonest causes of death in these years are coronary heart disease, cerebrovascular disease, cancer, pneumonia and bronchitis. Among males the predominant cancer is in the lung, among females in the breast. The trends are irregular. Cancer of the lung is increasing in males, though at a slackening rate. The number of reported cases of cancer of the breast has been increasing very slowly in recent years in these age groups ; but this may be due to better cancer registration and diagnosis. Coronary heart disease has been increasing, but mortality from cerebrovascular disease and bronchitis has remained fairly steady. The total volume of cardiovascular disease has been increasing slowly in males but decreasing slightly in females. Cancer of the cervix has been increasing very slightly. Cancer of the stomach has been decreasing slowly in both sexes.

Older ages (65 + years)

The patterns of middle life are broadly continued at older ages but the inadequacies of certification and the complexity of the death entries make analysis more uncertain. As age advances the discrepancies between male and female mortality remain but are less marked. On the whole male mortality is now stationary while that of females continues a slow decline.

SUMMARY

The general picture which one derives from the study of mortality from all ages has been noted by a number of authors, e.g. McKeown [2] and Morris.[3] There has been a steady decline in mortality, mainly from communicable diseases, particularly in early and middle life, and, in later life, an increase in the diseases of obscure aetiology, such as vascular disease and cancer. There has also been an increase in disease whose determination might be called 'social': suicide and road accidents among young men form an obvious group of this kind; but, from the preventive point of view, lung cancer too is almost a purely social problem. It also seems clear that the gap between the social classes in some forms of mortality represents a source of preventable deaths which can best be attacked by sociological rather than medical methods. It is particularly interesting for example that those who take the most advantage of screening programmes for cancer of the cervix are often those in the lowest risk groups. A similar situation may exist in the case of infant mortality, where some of the social and geographical variations could be due to failure to exploit the services that are available.

Heart disease is being dealt with by Professor Morris so I will not enlarge on it, but it presents something of an enigma. A very large part of the apparent increase can be ascribed to ageing of the population, and a change in fashion of diagnosis probably accounts for a good deal of the apparent rise in coronary thrombosis. The fairly stationary state of mortality from cerebrovascular accidents makes me feel that, at the older ages, there has not been a true increase in all forms of heart diseases at all ages, though there may have been in middle life. It is also worth noting that there has been a steady increase in deaths due to thrombotic phenomena at most ages and in both men and women.

Cancer of course still remains a mystery in many cases. Both in this and in heart disease the effect of natural selection in pushing causes of death into the non-reproductive ages is probably at work and the ultimate roots are probably very deep seated and difficult to affect by medical intervention.

Social factors of a preventable kind are, however, at work here also. For example, cancer of the stomach, which has a very marked social gradient, has been declining for many years and some of this decline is probably due to the general improvement in the standard of living.

The control of infectious disease is now very far advanced, but there may be opportunities for further advances in the control of virus diseases. The recurrent epidemics of acute respiratory disease and influenza are reflected in the death rates at all ages and are potentially controllable by vaccines and by the discovery of anti-viral agents. Much of the mortality from this cause is a terminal phenomenon in patients weakened by other diseases, but the number of lives that might be saved or prolonged is still large in absolute numbers.

I have given here only a very brief sketch of the current position and possible future trends in mortality but I hope that it will form a sufficient background for the more detailed papers in this symposium.

DISCUSSION

A speaker suggested that there had in fact been a significant trend in bronchitis mortality figures, with a decrease in female mortality very similar to the changes in the sex ratio shown in deaths from lung cancer. The figures for male mortality had remained constant because the greater effectiveness of the latest treatment had been counteracted by the increase in smoking. This improved trend showed in females, who did not smoke so much, but was hidden in males by an increase in smoking. Dr SPICER agreed with this, saying that it explained why the introduction of broad spectrum antibiotics had not had a greater effect on the male mortality rates; the effectiveness of treatment could have been counteracted by the rise in the incidence of mortality due to smoking.

REFERENCES

1. CAMPBELL, H. 1965. *Changes in Mortality Trends, England and Wales, 1931–61.* National Center for Health Statistics Analytical Studies No. 3. Washington.
2. McKEOWN, T. 1964. The next forty years in public health. *Popul. Stud.* 17, 269.
4. MORRIS, J. 1962. *The Uses of Epidemiology.* London and Edinburgh. Livingstone.

'CORONARY THROMBOSIS'
PROGRESS AND PROSPECTS

J. N. MORRIS

M.R.C. Social Medicine Research Unit, London Hospital

ISCHAEMIC heart disease, or ' coronary thrombosis ', is responsible for about 30 per cent of deaths of men in middle age in
this country, so for about 30 per cent of widowhood at its most
miserable, and about 30 per cent of widows' pensions. Adding
the cost of these to sickness benefit and expenditure on hospital
admissions yields an estimate of £50 million p.a. currently in
' social ' costs. Beyond that, there is a loss of income of something like £200 million p.a. to the families concerned. Ischaemic
heart disease (IHD) emerged from obscurity during the early
years of this century, and particularly after the first world war ;
it has been increasing, we can estimate, so that one in five men
of this country now develop the clinical condition during
middle age—this is the ' modern epidemic '.[7] It is not possible
at present to make any valid estimates of the numbers of men
affected in old age. Similarly less is known about the situation in women ; though evidently it is less of a menace among
them, until old age anyhow.

Standard of living

Ischaemic heart disease is uncommon in low-income populations. The classical study is that of the late Dr Bronte-Stewart
and his colleagues in Cape Town, where three populations
with very different standards live side by side.[1] The experience
of IHD among the Europeans resembled that of the highly
developed countries of the West ; the ' Cape Coloured ' men
had substantially less ; among the wretched Bantu the disease
was quite rare. With social and economic advance, however,
the disease appears to strike rather quickly and, not untypically,
countries like Yugoslavia and India are keenly interested in it.
Among ' western ' countries themselves, there are substantial

and persistent differences (Table I). In Sweden, with the highest standard of living in Europe, the disease obviously is a far smaller burden than in the UK, and this is one basis of

TABLE I

MORTALITY AND THE STANDARD OF LIVING
1961

(Rates per 1000)

COUNTRY AND RANK	ALL CAUSES 45–64 yrs.		ISCHAEMIC HEART DISEASE 45–54 yrs.	
	M	F	M	F
Kuwait	?		?	
Qatar	?		?	
USA	15·0	7·9	3·4	0·74
Canada	12·3	6·6	3·0	0·56
Sweden	9·1	6·0	1·0	0·20
Switzerland	12·0	6·6	1·0	0·16
Luxembourg	?		?	
Australia	12·9	6·9	2·7	0·60
New Zealand	12·5	7·5	2·6	0·62
UK	14·1	7·6	2·1	0·36

WHO/UNO/*Financial Times*

the hope that we might eventually be able to control it without scrapping western civilization.

Blood lipids and diet

I now briefly summarize some results of modern epidemiological research,[4, 11] beginning with the important blood cholesterol level. Characteristically, the Europeans in Cape Town had far higher levels than the Bantu men of the same age, the Cape Coloured being between. Such differences between populations and countries have been paralleled in one of the main findings of modern prospective studies. They show that healthy men free of ischaemic heart disease whose blood cholesterol level is high are more likely to develop the disease in the future—the ' incidence '—than comparable men with lower levels. Figure 1 is typical of these results.[8] The low average level of blood cholesterol (and of ischaemic heart disease) in low-income *populations* is commonly ascribed to lack of fat in their diet.[6]   There is no evidence, however, that individual differences of diet *within* high-income populations are related to the wide range of individual blood lipid levels found among them.   In Table II, our own data on this point

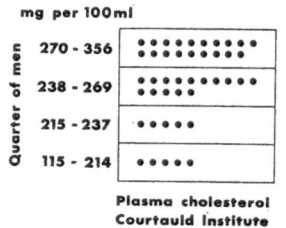

mg per 100ml

Plasma cholesterol
Courtauld Institute
of Biochemistry

FIGURE 1

Incidence—' first clinical attacks '— of ischaemic heart disease in sample of middle-aged London busmen during five years by plasma cholesterol level at initial examination.

The men have been divided into four quarters.   Each dot represents one new case.   607 men.

are given for two of the main nutritional factors currently under suspicion; for sugar a more relevant lipid fraction is quoted. In the case of fat, it may be that the entire population is above a level where intake is related directly and simply to blood cholesterol levels.   More generally, there may be a nutritional ' threshold ', above which the disease begins to be common,

TABLE II

INDIVIDUAL DIET AND BLOOD LIPID LEVELS IN SAMPLES OF BANK MEN

FATTY ACIDS RATIO $\frac{\text{Saturated}}{\text{Polyunsaturated}}$	PLASMA CHOLESTEROL mg. per 100 ml. Mean	DIETARY SUGAR g. per day	SF 20-400 LIPOPROTEIN mg. per 100 ml. Mean
4·0 to 5·9	241	9 to 78	136
5·9 to 6·7	228	79 to 109	165
6·7 to 7·8	242	110 to 141	162
7·8 to 11·0	239	142 to 206	154
Mean 6·9	Mean 237	Mean 109	Mean 154

The men weighed what they ate for one or two weeks.

There were 24 or 25 men in each diet-cholesterol group, and 18 or 19 in each diet—Sf20-400 group.   Ages 40–55 years.

Ranges of lipid levels in each group were wide, and similar.

and this threshold may well be at rather low levels of food-intake.

Prediction by the blood pressure

The other main result of these prospective studies is less of a discovery—that men with high casual blood pressure and, again, showing no evidence of ischaemic heart disease, are more likely to develop it for the first time (and other cardiovascular disease also) than comparable men with lower blood pressure levels. In our own studies (Figure 2) the systolic pressure was a somewhat better predictor of the disease than the diastolic. The men have again been divided into four equal quarters, on this occasion by their casual BP level. Those in the top quarter of the distribution may conveniently be labelled

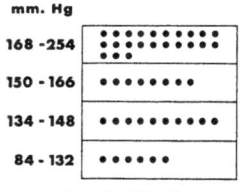

FIGURE 2

The incidence, by casual systolic blood pressure level. 664 busmen.

' high ' or ' hypertensive ' ; most of them have systolic pressures of 168–199 mm Hg. Each dot again represents a new case of IHD.

Now, it is exceedingly interesting that, although levels of blood cholesterol and of blood pressure are both related to age, to family history of cardiovascular disease, to obesity, etc., the correlation between these levels is low—in our data +0·28. That is to say, most busmen with hypercholesterolaemia did not have high blood pressure, etc. In consequence, as Figure 3 shows, men who have high levels of one or the other account for most of the disease—for the modern epidemic of IHD. The individual risk of new ischaemic heart disease among these busmen having a high level of either—the vulnerable 40 per cent— is about five times as great as among the 60 per cent of men with lower levels of both. The incidence among this 60 per cent, indeed, can scarcely be described as a major issue in our Public Health.

In search of causes

High levels of blood cholesterol and blood pressure may be called 'precursors' of the disease : they are evidence of disturbed metabolism and haemodynamics associated with a considerably increased risk of future ischaemic heart disease ; they are stages on the way, it may be said.[7] Progressing further into possible causes of these high precursor levels and of the disease itself, *causes* in inheritance and the mode of life, there is encouraging progress to report from the intensive researches of the last twenty years. I will not refer further to diet despite its importance.

FIGURE 3

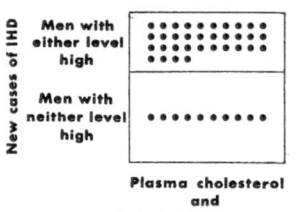

Men with either level high

Men with neither level high

New cases of IHD

Plasma cholesterol and Systolic blood pressure

The incidence of ischaemic heart disease in sample of middle-aged London busmen during five years. 607 busmen.

Upper part : Men with either plasma cholesterol or casual SBP in high quarter of distribution.

Lower part : Men with neither level in high quarter.

Behaviour pattern

A San Francisco group of investigators have now shown striking differences of incidence in relation to psycho-social factors. The 'Type A' men whom they described (Table III)

TABLE III

INCIDENCE OF ISCHAEMIC HEART DISEASE IN SAMPLE OF CALIFORNIAN MEN[10] Ages 50–59

Rates per 1000 man-years

TYPE A BEHAVIOUR PATTERN		TYPE B BEHAVIOUR PATTERN	
ALL MEN	24	ALL MEN	12
Hypertensives	59	Hypertensives	17
Others	18	Others	11

are busy with multiple involvements 'vocational and avocational' in a competitive society, and their lives are dominated by time schedules. Men of Type B live at a slower pace and are blessed with more equanimity. These investigators also, of course, found the disease more common among the hypertensive men; very promising is the suggestion that high blood pressure may be important only in the presence of Type A behaviour pattern.

Sedentary living

Figure 4 is typical of an observation now made in many countries relating occupation to incidence; it is postulated

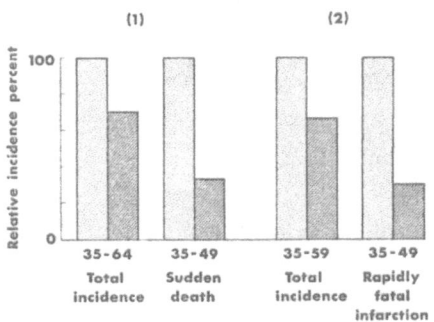

FIGURE 4

Relative incidence of ischaemic heart disease in London busmen and in government workers : [7]
(1) *Dotted column*: Drivers; *Hatched column*: Conductors.
(2) *Dotted column*: Clerks; *Hatched column*: Postmen.

that the physical activity/inactivity involved in jobs is responsible for the differences in ischaemic heart disease. (The only study where physically active workers consistently have as much IHD as sedentary men is the one made in Los Angeles.[2]) We can take this observation further. The physically active bus conductors have lower blood lipid levels than corresponding drivers. Table IV(A) shows the findings for beta-lipoprotein cholesterol, believed to be the relevant fraction in the

F

TABLE IV

OCCUPATION; OBESITY; AND BLOOD LIPIDS IN LONDON BUSMEN
(mg. per 100 ml.)

| | (A) | | | | | (B) | |
| | AGE | | | | | | AGE | |
	35–39	40–49	50–59	60–64			40–49	50–59
BETA-LIPO-PROTEIN CHOLESTEROL						BETA-LIPO-PROTEIN CHOLESTEROL		
Conductors	173	185	190	189		Conductors		
Drivers	195	200	202	196		Thin	156	170
SF 20–400						Fat	190	202
Conductors	116	119	101	96		Drivers		
Drivers	170	164	156	122		Thin	178	187
						Fat	220	207

550 men.   354 men.
Courtauld Institute of Biochemistry.

plasma cholesterol, and for the very low density lipoproteins in which there is much interest to-day. The conductors are also thinner than the drivers on average, and there are far fewer obese conductors.[7] Obesity also is associated with high lipid levels, but this does not explain the occupational differences found. Table IV(B) summarizes the occupational effect within grades of skinfold-thickness: the range is now 156 to 220 mg. per 100 ml. at ages 40–49 instead of 185 to 200, and 170 to 207 mg. per 100 ml. at 50–59 years instead of 190 to 202. The occupational-physical activity factor is not merely a matter of leanness/fatness.

Bus conductors also have lower blood-pressure levels on average, after fifty years of age. At the *same* levels of blood pressure, moreover, the conductors seem to have a lower incidence of IHD than the drivers.

Cigarette smoking

Prospective studies, as in Table V, are quite clear about the dangers of cigarette smoking and the immunity of pipe and cigar smokers, reward for virtue.

TABLE V

SMOKING AND THE RISK OF ISCHAEMIC HEART DISEASE

Incidence during 12 years According to Smoking Status
on Entry to Study at Ages 30–59

Men

Framingham

MORBIDITY RATIOS FOR IHD *

ALL MEN	100
Cigar and pipe smokers	41
Ex-Smokers	60
Non-Smokers	82
Mild/moderate cigarette smokers	110
Heavy cigarette smokers	155

* Excluding angina pectoris.

Family history

Several studies have shown a higher frequency of the
disease when first-degree relatives are already affected, and I
hope this important aspect will be dealt with by one of the
experts contributing to this symposium.

PROSPECTS

I have described briefly and, I fear, rather categorically,
some of the main findings of modern investigation. In terms
of controlling the disease, there now are two openings. First,
we must continue trying to teach people to lead more healthy
lives, adapting health-educational techniques to modern conditions. As Sydenham put it, " the tranquillity of the mind
is to be established by all possible means ". There is already
enough evidence to encourage men to take more exercise :
obesity-high blood lipids/high blood pressure—' coronary
thrombosis ' and diabetes reflect an ' epidemic constitution '
of our time with its technology and affluence ; mental over-
stimulation and physical sloth ; over-eating, cigarette smoking, etc. Systematic efforts have been started to educate the
public on the dangers of cigarette smoking. It is early to say ;

but the most recent figures do provide a glimmer of hope—cigarette smoking does seem to have fallen a little in the population since 1962 when the Royal College of Physicians brought the issue forcibly before the people (Table VI). Manifestly, however, the barrage of facts is not getting the necessary response.

TABLE VI

RECENT HISTORY OF CIGARETTE SMOKING

Britain

	ANNUAL, LBS. All Adults		WEEKLY NOS. 16–19 yrs.	
	M	F	M	F
1961	9·4	3·6	57	25
1962	8·8	3·5	65	30
1963	8·8	3·7	57	26
1964	8·6	3·6	57	28
1965	8·0	3·5	50	26

Tobacco Research Council.

The brighter prospect now opening up is that men at high risk of developing the disease can be identified early, health-educational measures addressed to them and, what is more to the point, personal *treatment* offered in the hope of reducing their risk. This raises the question, about as important as any in modern medicine, how mild and moderate hypertension and hypercholesterolaemia can be effectively controlled over periods of years, and whether this control will lower the incidence of ischaemic heart disease. On present (or conceivable) availability of treatment, such an approach is scarcely possible for everyone concerned: even with perfect prediction of middle-aged individuals at particular risk for IHD that would mean treating one-fifth of the male population. The principle, however, may be applicable to narrower groups. For example, it should be possible eventually to define, by family history and other risk factors, men who are particularly likely to develop the disease in early middle age, or to develop

it very severely, or very soon. Both dietary and pharmacological methods are now being employed in trials to lower high
blood cholesterol levels, and tentative results with diet are
encouraging. Meanwhile, little attempt is being made in
large-scale trials to lower hypertension in otherwise healthy
men, and it is doubtful whether any of the agents at present
available are sufficiently safe and effective (this is a different
question, of course, from the obvious clinical-individual need
to reduce very high levels of pressure). We look expectantly to
the pharmacologists.

Better methods of identifying susceptible men and, then,
of controlling their susceptibility may not be too far away;
and with this glimpse of tomorrow's preventive medicine I
close.

The studies reported are a team effort of the Social Medicine
Research Unit, MRC, and I am grateful to my colleagues
whose particular contributions are quoted, in particular to
Miss J. W. Marr, Dr Aubrey Kagan and Dr D. C. Pattison,
Dr J. A. Heady and Mr M. J. Gardner. The Unit is greatly
obliged to Dr G. L. Mills, Courtauld Institute of Biochemistry,
Middlesex Hospital, for making all the blood lipid estimations.

DISCUSSION

In reply to a question on blood pressure, PROFESSOR MORRIS
said that although in other studies the diastolic pressure had
been found to be a slightly better predictor, he and his co-
workers had not found it so good as the systolic, and various
combinations of blood pressure at rest and after stimulation
of different kinds were no better. The importance of blood
pressure being labile was very interesting : in quite a number
of men the pressure went down during the course of the
examination when they relaxed on a couch and chatted.

The Chairman recalled that Professor Morris had been an
important member of the committee which produced the
Royal College of Physicians Report in 1962, and asked what

the figures had been for the advertising of cigarettes since the Report was published. Professor Morris said that the figures were very unsatisfactory; they had risen steeply.

Professor Morris was then asked to enlarge on the findings in South Africa. He said that the mortality among middle-aged European males tended to be rather higher than the mortality in this country and to approximate more to that in America. The mortality among ' Cape coloureds ' was about half as much. Among the Bantu there was no discernible coronary heart disease; the rate was so small as to be scarcely worth putting down: their average blood cholesterol was about 60 per cent of the average in Europeans.

REFERENCES

1. BRONTE-STEWART, B., KEYS, A., BROCK, J. F., MOODIE, A. D., KEYS, M. H. and ANTONIS, A. 1955. Serum-cholesterol, diet, and coronary heart-disease. *Lancet* ii, 1103.
2. CHAPMAN, J. M. and MASSEY, F. J. 1964. The interrelationship of serum cholesterol, hypertension, body weight, and risk of coronary disease: Results during first ten years' follow-up in the Los Angeles heart study. *J. chron. Dis.* 17, 933.
3. CHRISTAKIS, G., RINZLER, S. H., ARCHER, M., WINSLOW, G., JAMPEL, S., STEPHENSON, J., FRIEDMAN, G., FEIN, H., KRAUS, A. and JAMES, G. 1966. A dietary approach to the prevention of coronary heart disease—a seven-year report. *Amer. J. publ. Hlth.* 56, 299.
4. EPSTEIN, F. H. 1965. The epidemiology of coronary heart disease—a review. *J. chron. Dis.* 18, 735.
5. KANNEL, W. B., DAWBER, T. R. and McNAMARA, P. M. 1966. Detection of the coronary-prone adult: The Framingham study. *J. Iowa Med. Soc.* 56, 26.
6. KEYS, A. 1962. Diet and coronary heart disease throughout the world. *Cardiol. Prat.* 13, 225.
7. MORRIS, J. N. 1964. *Uses of Epidemiology.* Edinburgh and London. Livingstone.
8. MORRIS, J. N., KAGAN, A., PATTISON, D. C., GARDNER, M. J. and RAFFLE, P. A. B. 1966. Incidence and prediction of ischaemic heart-disease in London busmen. *Lancet* ii, 553.
9. MORRIS, J. N., MARR, J. W., HEADY, J. A., MILLS, G. L. and PILKINGTON, T. R. E. 1963. Diet and plasma cholesterol in 99 bank men. *Brit. med. J.* i, 571.
10. ROSENMAN, R. H., FRIEDMAN, M., STRAUS, R., WURM, M., JENKINS,

C. D., Messinger, H. B., Kositchek, R., Hahn, W. and Werthessen, N. T. 1966. Coronary heart disease in the western collaborative group study. *J. Amer. med. Assoc.* **195**, 86.

11. Stamler, J., Berkson, D. M., Lindberg, H. A., Hall, Y., Miller, W., Mojonnier, L., Levinson, M., Cohen, D. B. and Young, Q. D. 1966. Coronary risk factors. *Med. Clins. N. Amer.* **50**, 229.

12. Turpeinen, O., Karvonen, M. J. and Roine, P. 1966. Dietary prevention of coronary heart disease. *VIIth International Congress of Nutrition. Hamburg.*

ENVIRONMENTAL AND CONSTITUTIONAL FACTORS IN THE AETIOLOGY OF LUNG CANCER

C. M. Fletcher

Postgraduate Medical School of London

The magnitude of the problem presented by lung cancer is shown by the fact that in 1965 among men in England and Wales, 8 per cent of all deaths and 39 per cent of cancer deaths were due to the disease. In two respects it is a unique form of malignant disease.

First, its incidence has shown a tremendous increase during the past fifty years in every country for which reliable statistics are available.[38, 39] The absolute level of mortality varies greatly from country to country, but the universal increase, far greater in men than in women, is most impressive. There is little doubt that a real and steady increase, at least in men, must have begun in the first twenty years of this century in many countries. There is no doubt about the continuing increase in both sexes during the past twenty years.[45]

The second unique feature of lung cancer is the strength of the evidence that most cases are due to environmental factors. If we accept this evidence and act upon it we should be able to prevent a vast number of deaths, many of which are premature and painful. I propose to consider these environmental causes of lung cancer in order of their importance.

CIGARETTE SMOKING

The evidence that cigarette smoking is an important cause of lung cancer is, in summary :

1. Close quantitative relationship between cigarette consumption and lung cancer mortality has been found in at least thirty retrospective studies in nine different countries and in seven prospective studies in three countries.[30, 50b] These have

shown a remarkably consistent and direct association between the number of cigarettes smoked and the mortality from lung cancer. Only two surveys, one of autopsy material [41] and the other of tobacco employees,[7] have failed to confirm this association and both were faulty in design.[33, 50d] These surveys have consistently shown also that the association is with cigarette smoking and not with pipe or cigar smoking. In some of these surveys it has been shown that those who inhale have a higher mortality than those who do not.[15, 50c] It is the strength and consistency of the association between lung cancer and cigarette smoking that provides so powerful an argument in support of its being a case of cause and effect. The alternative possibility, that the association is due to some factor which is independently associated with both cigarette smoking and liability to lung cancer, is much less probable if only because there is so little evidence for any such factor.

2. In all the surveys it has been shown that ex-smokers have a lower mortality than those who continue to smoke. The most detailed study of the effects of stopping smoking has been Doll and Hill's survey of British doctors,[15] which showed that those who give up smoking exhibit a rapid decline in lung cancer mortality as compared with those who continue to smoke. A consequence of this has been that since at least 30 per cent of doctors who formerly smoked cigarettes have stopped during the past ten years, there has been over this period a 7 per cent decline in lung cancer mortality among doctors, whereas in the general population mortality has risen by more than 22 per cent.

3. Auerbach and his colleagues [1a, b] have shown that pathological changes in the bronchial epithelium which many pathologists consider may be pre-cancerous, are quantitatively related to cigarette smoking and are rare in non-smokers, ex-smokers and smokers of pipes and cigars.

4. Cigarette smoke contains several substances capable of producing carcinoma in experimental animals.[37, 50a] Although concentration of these carcinogenic substances in cigarette smoke has been thought to be too small to produce the attributed carcinogenic effect,[8, 60] this objection is not based on strictly comparability evidence, and smoke also contains a

number of co-carcinogenic substances [50a] which could enhance the effects of these and other carcinogens.[40] While no one has yet produced a true bronchial carcinoma in animals by short-term exposure to cigarette smoke, it must be remembered that animal experience is not always relevant to man. There is no doubt, for instance, that arsenic can be carcinogenic in man, but it has never produced a carcinoma in any other animal.[28]

The existence of a close association between cigarette smoking and lung cancer is now universally accepted. The conclusion that the association is one of cause and effect has been persistently challenged by only a few people, some of whom either have an interest in the sale or manufacture of cigarettes, or are so manifestly biased that despite their denials it appears that they must be trying to avoid a conclusion which for some reason is unpalatable for them. While the evidence is strong, it does not amount to proof, nor could proof be obtained except by wholly unethical human experiments. To counter the causative hypothesis requires evidence with which it is incompatible. The only important facts that appear at first sight difficult to reconcile with it are :

1. The relationship between cigarette smoking and lung cancer is quantitatively different in different areas of the same country [12, 25, 26, 47] and in different countries; and emigrants from the United Kingdom, with its high mortality, to other countries with lower rates maintain a relatively high mortality.[10, 11, 24] These facts provide the main evidence for an effect of air pollution ; but, since a quantitative relationship between cigarette consumption and lung cancer is universal, they do not counter the causation hypothesis, but merely show that other factors, which modify the effects of cigarette smoke on the lung, are involved. Apart from air pollution, for instance, differences in detailed smoking habits such as the amount of each cigarette that is smoked, the rate of smoking, the amount of inhalation and the type of tobacco that is smoked, may all affect the exposure of the lung and the chemical and physical components of the cigarette smoke which reach it.

2. Although pipe and cigar smoke contains more carcinogenic substances than cigarette smoke,[50a] smokers of pipes and

cigars show only a small increase of lung cancer mortality. The effect of inhalation may be important here, for few pipe and cigar smokers inhale. This explanation is consistent with the fact that there is a similar increase in mortality from oral cancer in smokers of all kinds.[50c]

The comparative innocuousness of smoking pipes and cigars does not counter the evidence that cigarette smoking is a cause of lung cancer. It only throws some doubt on the relevance of laboratory measurements of the carcinogenic content of the smoke.

3. During the past twenty years cigarette consumption has been increasing more rapidly among women than among men in this country, but lung cancer mortality has increased more rapidly among men. This is not surprising in view of the long latent period of carcinogenesis and since the increase in female cigarette smoking has occurred predominantly in younger women who are only just approaching the age of increased liability to cancer. If we compare men and women under the age of fifty, the ratio of average lifetime cigarette consumption is about 3 : 1, which is close to the 4 : 1 ratio of lung cancer mortality at this age.[55]

4. Cigarette smoking is associated with increased mortality from a large number of diseases. It has been argued [2] that it is impossible for one cause to have so many effects, and that the association must therefore be due to a general tendency for the sort of people who smoke to ' live faster ' and thus die more readily than non-smokers. This hypothesis fails to account for the far greater mortality ratio of smokers to non-smokers for lung cancer than for any other disease. Cigarette smoke is a very complex substance and, as Cornfield *et al.*[9] have pointed out, " there is nothing contradictory or inconsistent in the suggestion that one agent can be responsible for more than one disease. . . . The great fog of London in 1952 increased the death rate from a number of causes, particularly respiratory and coronary disease, but no one has given this as a reason for doubting the causal role of the Fog. . . . A universe in which cause and effect always have a one-to-one correspondence with each other would be easier to understand, but it obviously is not the kind we inhabit."

The evidence that cigarette smoking is an important—probably the most important—cause of lung cancer is so consistent, so persuasive and so free from valid objection, that it is unreasonable not to accept it until some convincing alternative explanation has been put forward. (I shall deal with the inadequacies of the constitutional hypothesis later.) The evidence supporting this conclusion is incomplete in many important particulars and much more research is needed to fill these gaps; but it is certainly strong enough to justify basing preventive measures upon it.

AIR POLLUTION

In relation to lung cancer, air pollution is usually taken in a non-specific sense to indicate the general pollution of the air, particularly of our cities and their surroundings, by the products of modern industrial activity, in particular coal smoke, road dust, motor vehicle exhaust fumes and industrial effluents. It is important to realize how ill-defined and, indeed nebulous, the concept is. While we can characterize most people quantitatively according to their smoking habits, they are difficult to characterize precisely in terms of their inconstant exposure to this qualitative type of ' air pollution ' and this is a great handicap to accurate epidemiological research. The hypothesis that air pollution may increase lung cancer mortality is mainly based on three facts.

1. In many countries [58] there is a gradient of lung cancer mortality from large towns through smaller towns to rural areas. This difference is not due to differences in current cigarette consumption as between urban and rural dwellers. In each type of area a gradient with cigarette smoking is observed.[12, 25, 26, 47] There is some evidence that the apparent effects of urban environment and cigarette smoking may be more than merely additive.[25, 26]

2. Emigrants from the United Kingdom, with its generally high levels of air pollution, to other countries with lower levels have a higher lung cancer mortality rate than the indigenous population.[10, 11] Emigrants from Norway to the USA contrariwise maintain a lower mortality than the indigenous.[24]

3. The products of combustion of coal and petrol contain carcinogenic substances similar to those found in tobacco smoke.[29, 54] The concentration of these substances is much greater in urban air than in rural air. Stocks [48] has shown a high correlation in a selected group of towns in North-west England between measurements of the smoke content of the air and lung cancer mortality. Smoke concentration and carcinogenic concentration run parallel.[49]

While these facts provide suggestive evidence of an important effect of air pollution, not only are there several important contradictions in the evidence, but there are other facts that suggest that air pollution cannot be an important cause of lung cancer.

1. In most countries, and certainly in the United Kingdom, pollution by coal smoke has been decreasing [56] while lung cancer mortality has been increasing, and so coal smoke cannot be the cause of the modern lung cancer epidemic. Pollution by petrol exhausts has increased on a time-scale similar to that for increasing cigarette consumption, but the increase of pollution by diesel exhausts began after the steep rise in the lung cancer incidence had started.[21] There is no increased mortality from lung cancer among long-distance lorry-drivers whose exposure to petrol exhausts should be high,[13] nor is there any difference in lung cancer mortality between predominantly rural and urban bus drivers.[43] Air pollution by petrol exhaust is very high in Los Angeles, but lung cancer is rare in the non-smoking sect of Seventh-Day Adventists,[59] many of whom live in that area.

2. The close correlation between measurements of air pollution and lung cancer observed by Stocks [48] is not confirmed when a wider sample of urban areas in the United Kingdom is considered.[57] Other attempts to correlate lung cancer mortality with pollution measurements have produced inconsistent results.[5, 19]

3. The urban/rural gradient of lung cancer mortality is as steep in Scandinavian countries, where levels of pollution are very low, as it is in industrialized England.[6]

4. Men and women have approximately equal exposure to urban air but a very different lung cancer mortality.

5. Lung cancer mortality is very high in Finland, where air pollution is low while cigarette consumption has been very high for many years. In Japan with very intense air pollution lung cancer mortality is low.[44]

6. Gas workers are exposed to concentrations of carcinogenic substances (to which the effects of air pollution have been attributed) that are a hundred times greater than the concentration found in urban air,[34] and yet their lung cancer mortality is only twice that of other men in urban areas.[16]

Doubtless, there is some urban factor which, even in the absence of cigarette smoking, has an effect on the incidence of lung cancer and which appears to increase the hazards of cigarette smoking considerably. The nature of this urban factor remains uncertain. Some studies have shown a close correlation of lung cancer mortality with indices of overcrowding rather than with indices of pollution.[19]

OCCUPATION

There is no doubt that working in certain occupations increases the risk of lung cancer. The subject has been reviewed by Doll.[14] The occupations incriminated are:

Definite risk

        Asbestos mining and manufacture
        Chromatin manufacture
        Gas production
        Nickel refining
        Mining of radio-active materials

Suspected risk

        Arsenic handling
        Manufacture and mining of iron
        Hard rock mining

The number of men employed in these industries is too small for these hazards to make a significant contribution to total cancer incidence. The theoretical importance of these occupational risks is their demonstration of the fact that long-term exposure to a variety of substances, not all of which would be thought to be carcinogenic, may in fact produce lung cancer.

CONSTITUTION

The hypothesis that constitution is relevant to the aetiology of lung cancer is thus proposed by Eysenck,[17] one of its main proponents : " What is maintained, in effect, is that there are certain types of people who smoke cigarettes ; that this type of person has acquired his particular personality through hereditary causes and that this particular type of person is also more likely to develop cancer."

There are two aspects of this hypothesis. First, on the smoking side, it has been well established from twin studies that there is a genetic or constitutional factor in smoking habits. Four separate surveys have shown that the smoking habits of identical twins are more concordant than those of non-identical twins.[20, 23, 42, 53] Smokers have been shown to differ from non-smokers also in their physical characteristics,[46, 51] but there is little difference in this respect between those who smoke more and those who smoke fewer cigarettes. There are many psychological differences between smokers and non-smokers.[36, 50f] Perhaps the clearest differentiation has been that reported by Eysenck et al.[18] who reported a smooth gradient, from introverted non-smokers and pipe smokers to a steadily increasing extraversion of smokers with increasing consumption of cigarettes. This general trend has been confirmed in other surveys.[50f]

The other aspect of the hypothesis is that there is an inherited liability to lung cancer which is closely and quantitatively associated with the proneness to smoke cigarettes. The evidence in support of this proposition is scanty. Association between lung cancer and specific personality types has been studied only by Kissen and Eysenck,[32] who report that compared with patients with non-malignant chest diseases there is a tendency for lung cancer patients (but only those free from psychosomatic disorders) to show personality characteristics not unlike those of cigarette smokers. Eysenck [17] states dogmatically, without quoting evidence, that this personality association is not just due to lung cancer patients being heavier smokers than the controls and thus inevitably having the psychological characteristics of heavy smokers—which seems to me to remain a more probable explanation of the findings. Kissen [31] emphasizes the low ' neuroticism ' score of the cancer patients and has calculated that men with a low score for neuroticism may have a lung cancer mortality five times greater than those with high scores. These findings are interesting and challenging, although at present they are based on meagre evidence. The hypothesis that cancer may be determined by personality characteristics is one I find difficult to accept, but there is some evidence that it may be true for a number of different types of cancer.[35]   Indeed, apart from cancer of the stomach and colon, of the breast and retinoblastoma, there is little or no evidence of any genetic determination of cancer.[6]   Twin studies, hitherto on small numbers, have provided no evidence of a genetic basis for lung cancer, but some family aggregation has been reported.[52]

The theoretical basis of the hypothesis that a common constitutional factor is responsible both for an increased desire for cigarette smoking and liability to lung cancer must be complex.   It demands a close quantitative relationship not only between cancer liability and the amount smoked, but within each quantitative group a relationship with the age at which smoking is started.   This close genetic gradation would also have to include the reduced liability of ex-smokers to smoke and to get lung cancer, and for this liability to lung cancer to decrease rapidly after the time at which smoking is stopped.[15]   The imagination boggles at the vast number of genotypes demanded by such a complex hypothesis, which seems to me to be so very much less likely than the hypothesis that exposure to cigarette smoke is a cause of lung cancer. Moreover this improbable hypothesis fails to account for the following important facts :

1. *The rising mortality from lung cancer.*   Some advocates of the constitutional hypothesis [4] seek to explain it on the grounds that, with declining mortality from other lung diseases such as tuberculosis and pneumonia, those with the appropriate cancer-prone constitution now survive to die of lung cancer.

This explanation is invalidated, however, by the fact that the increase in lung cancer is greater among males, while the decline in other chronic respiratory diseases has been approximately equal in the two sexes. Others [17] blame air pollution, ignoring the contrary evidence summarized above.

2. *The fall in lung cancer mortality among doctors.* Most of the doctors who have stopped smoking in the past ten years are not like ex-smokers in general who on the constitutional hypothesis may have inherited a low desire for smoking and a low liability to lung cancer. They are men who, despite a desire for cigarettes, have given up the habit because of the evidence of its danger. They appear to be reaping a reward which, by the constitutional hypothesis, should not be theirs.

At present such preventive efforts as are being made to contain the flood of cigarette smoking are meeting with only meagre success.[22] There has been a recent decline in the proportion of cigarette smokers among elderly men, but women and the younger generation appear to be little affected. It may be wasteful to attempt to dissuade all cigarette smokers from their solace, for only a minority of them suffer from lung cancer and other diseases related to smoking, and it might be easier if we could concentrate our efforts on the susceptible subjects. We cannot at present recognize any constitutional trait that is a clear danger sign. The suggestion that smokers who have a cough are more prone to cancer has not yet been shown to have preventive value.[3]

If the constitutional hypothesis were true and if we could recognize those with a special liability to lung cancer, we might perhaps consider eugenic measures to breed out this undesirable trait from the human race. If we did this we would simultaneously remove the desire to smoke cigarettes and free the surviving non-smokers from the odour of stale tobacco smoke which now pollutes their environment. But we would also remove many of the most adventurous and amusing people whose company we enjoy. The constitutional hypothesis, however, is so unlikely to be true that we need not worry about this extravaganza. What we have to do is to devote far greater efforts to removing or modifying the environmental causes of lung cancer, chief among which is the cigarette.

G

Replying to questions, DR FLETCHER said that the question of what determined whether a person would or would not give up smoking was one on which little research had been done. The Central Office of Information had recently carried out a survey concerned with motivation of smoking which suggested that about one-third of smokers were ' habit smokers ', that is to say they smoked chiefly on social occasions and when in company, whereas two-thirds admitted to wanting to smoke chiefly when they felt agitated, nervous or on edge, or when working. They seemed to be getting pharmacological benefit from their smoking and might thus be regarded as habituated or even addicted. In this connection it was interesting to note that anti-smoking clinics reported approximately 30 per cent success in dissuading patients from smoking and that about 30 per cent of smoking doctors gave up the habit between 1951 and 1958.[15] Perhaps ' habit smokers ' gave up smoking more easily than ' habituated smokers '.

As to which smokers did and which did not get cancer, it was quite likely that this was due to some random process. The alternative hypothesis was to suggest, as Eysenck and Kissen had done, that there was a cancer-prone group with a particular physiological make-up. It would be useful to be able to distinguish susceptible from non-susceptible smokers; but at present this was not possible.

The increase of smoking among women was chiefly occur-ring in the younger age groups; there appeared to be very heavy social pressure on young girls to smoke to be ' with it '.

The question of whether it was possible to produce a safe cigarette was a most important one. If a cigarette that might be expected to be safer were produced, it would take at least twenty years to discover whether the expectation was justified and this would require a large number of people to smoke these cigarettes exclusively. So far as bronchitis was concerned, it might be possible to show quite quickly that a modified cigarette produced less cough and sputum.

A steep increase in the price of cigarettes would probably be the most effective preventive measure. The high price of spirits had made an important contribution to the decline of alcoholism. A big increase in cigarette prices might, however, encourage stealing and smuggling of cigarettes and tobacco. The Chairman pointed out that a big increase in price might also make big inroads into the family income where both husband and wife were addicted smokers, and the level of nutrition of the family might suffer.

REFERENCES

1a. AUERBACH, O., STOUT, A. D., HAMMOND, E. C. and GARFINKEL, L. 1961. Changes in bronchial epithelium in relation to cigarette smoking and in relation to lung cancer. *New Engl. J. Med.* **265**, 253.

1b. AUERBACH, O., STOUT, A. D., HAMMOND, E. C. and GARFINKEL, L. 1962. Bronchial epithelium in former smokers. *New Engl. J. Med.* **267**, 119.

2. BERKSON, J. 1958. Smoking and lung cancer: some observations on two recent reports. *J. Amer. statist. Ass.* **53**, 28.

3. British Medical Journal. 1966. Leading article—Cough and Cancer. *Brit. med. J.* ii, 903.

4. BROWNLEE, K. A. 1965. A review of *Smoking and Health. J. Amer. statist Ass.* September, 722.

5. BUCK, S. F. and BROWN, D. A. 1964. Mortality from lung cancer in relation to smoke and sulphur dioxide concentration, population density and social index. *Tobacco Research Council London Research papers* No. 7.

6. CLEMMMESON, J. 1965. *Statistical Studies in the Aetiology of Malignant Neoplasms* 1. Review and results, p. 189. Copenhagen. Munksgaard.

7. COHEN, J. and HEIMANN, R. K. 1962. Heavy smokers with low mortality. *Industr. Med. Surg.* **31**, 115.

8. COOK, J. W. 1957. Chemical carcinogens and their significance. *Lancet* i, 333.

9. CORNFIELD, J., HAENSZEL, W., HAMMOND, E. C., LILIENFELD, A. M., SHIMKIN, M. B. and WYNDER, E. L. 1959. Smoking and lung cancer: recent evidence and a discussion of some questions. *J. nat. Cancer Inst.* **22**, 173.

10. DEAN, G. 1961. Lung cancer among white South Africans: report on a further study. *Brit. med. J.* ii, 1599.

11. DEAN, G. 1962. Lung cancer in Australia. *Med. J. Aust.* **1**, 1003.

12. DEAN, G. 1966. Lung cancer and bronchitis in Northern Ireland, 1960–2. *Brit. med. J.* i, 1506.

13. DOLL, R. 1953. Bronchial carcinoma: incidence and aetiology. *Brit. med. J.* ii, 521.

14. DOLL, R. 1959. Occupational lung cancer: a review. *Brit. J. industr. Med.* 16, 181.

15. DOLL, R. and HILL, A. B. 1964. Mortality in relation to smoking: ten years observations of British doctors. *Brit. med. J.* i, 1399.

16. DOLL, R., FISHER, R. E. W., GAMMON, E. J., GUNN, W., HUGHES, G. O., TYRER, F. H. and WILSON, W. 1965. Mortality of gasworkers, with special reference to cancers of the lung and bladder, chronic bronchitis and pneumoconiosis. *Brit. J. industr. Med.* 22, 1.

17. EYSENCK, H. J. 1965. *Smoking, Health and Personality.* London. Weidenfeld and Nicolson.

18. EYSENCK, H. J., TARRANT, M., WOOLF, M. and ENGLAND, L. 1960. Smoking and personality. *Brit. med. J.* i, 1456.

19. FAIRBAIRN, A. S. and REID, D. D. 1958. Air pollution and other local factors in respiratory disease. *Brit. J. prev. soc. Med.* 12, 94.

20. FISHER, R. A. 1958. Lung cancer and cigarettes. *Nature* 182, 108 and 596.

21. FLETCHER, C. M. 1965. Environmental factors in respiratory disease. In *Symposium on Advanced Medicine.* 243. (Ed. N. Compston.) London. Pitman Medical.

22. FLETCHER, C. M. 1965. Some recent advances in the prevention and treatment of chronic bronchitis and related disorders. *Proc. Roy. Soc. Med.* 58, 918.

23. FRIBERG, L., KAIJ, L., DENCKER, S. J. and JONSSON, E. 1959. Smoking habits of monozygotic and dizygotic twins. *Brit. med. J.* i, 1090.

24. HAENSZEL, W. 1961. Cancer mortality among the foreign-born in the United States. *J. nat. Cancer Inst.* 26, 37.

25. HAENSZEL, W., LOVELAND, D. B. and SIRKEN, M. G. 1962. Lung cancer mortality as related to residence and smoking histories. I. White males. *J. nat. Cancer Inst.* 28, 947.

26. HAENSZEL, W. and TAEUBER, K. E. 1964. Lung cancer mortality as related to residence and smoking histories. II. White females. *J. nat. Cancer Inst.* 32, 803.

27. HAMMOND, E. C. 1958. Lung cancer death rates in England and Wales compared with those in the U.S.A. *Brit. med. J.* ii, 649.

28. HILL, A. B. 1965. The environment and disease : association or causation. *Proc. Roy. Soc. Med.* 58, 295.

29. HUEPER, W. C., KOTIN, P., TABOR, E. C., PAYNE, W. W., FALK, H. and SAWICKI, E. 1962. Carcinogenic bioassays on air pollutants. *Arch. Path.* 74, 89.

30. JONES, D. L. 1966. An epidemiological study of certain aspects of lung cancer in New South Wales. *Med. J. Aust.* 1, 765 .

31. KISSEN, D. M. 1964. Personality and lung cancer. *Lancet* i, 216.

32. KISSEN, D. M. and EYSENCK, H. J. 1962. Personality in male lung cancer patients. *J. psychosom. Res.* 6, 123.

33. KOLLER, S. 1964. Bemerkungen zu der Arbeit von R. Poche, O. Mittman und O. Kneller. *Z. Krebsforsch.* 66, 187.

34. LAWTHER, P. J., COMMINS, B. T. and WALLER, R. E. 1965. A study of

the concentrations of polycyclic aromatic hydrocarbons in gasworks retort houses. *Brit. J. industr. Med.* **22**, 13.

35. LeShan, L. 1959. Psychological states as factors in the development of malignant disease : a critical review. *J. nat. Cancer Inst.* **22**, 1.

36. Lilienfeld, A. M. 1959. Emotional and other selected characteristics of cigarette smokers and non-smokers as related to epidemiological studies of lung cancer and other diseases. *J. nat. Cancer Inst.* **22**, 259.

37. Lindsey, A. J. 1962. Some observations upon the chemistry of tobacco smoke. In *Tobacco and Health* p. 21. (Ed. G. James and T. Rosenthal.) Springfield, Illinois. Charles C. Thomas.

38. Pascua, M. 1952. The evolution of mortality in Europe during the twentieth century : cancer mortality. *Epidem. vital. Statist. Rep.* **5**, 1.

39. Pascua, M. 1955. Increased mortality from cancer of the respiratory system. *Bull. Wld. Hlth. Org.* **12**, 687.

40. Pike, M. C. and Doll, R. 1965. Age at onset of lung cancer : significance in relation to effect of smoking. *Lancet* i, 665.

41. Poche, R., Mittman, O. and Kneller, O. 1964. Statische Untersuchungen über das Bronchial-karzinom in Nordrhein-Westfälen. *E. Krebsforsch.* **66**, 87 and 250.

42. Raaschou-Nielsen, E. 1960. Smoking habits in twins. *Danish med. Bull.* **7**, 82.

43. Raffle, P. A. B. 1957. The health of the worker. *Brit. J. industr. Med.* **14**, 73.

44. Sakabe, H. 1964. Air pollution in Japan. *Proc. Roy. Soc. Med.* **57**, 1005.

45. Segi, M. and Kurihara, M. 1963. Cancer mortality for selected sites in 24 countries. Graphic edition (Department of Public Health, Tohoku University School of Medicine, Sendai, Japan).

46. Seltzer, C. C. 1963. Morphologic constitution and smoking. *J. Amer. med. Ass.* **183**, 639.

47. Stocks, P. 1957. Cancer in North Wales and Liverpool region. *Brit. Emp. Cancer Camp., 35th Ann. Rpt.* Pt. II.

48. Stocks, P. 1960. On the relations between atmospheric pollution in urban and rural localities and mortality from cancer, bronchitis and pneumonia with particular reference to 3 : 4 benzopyrene, beryllium, molybdenum, vanadium and arsenic. *Brit. J. Cancer* **14**, 397.

49. Stocks, P., Commins, B. T. and Aubrey, K. V. 1961. A study of polycyclic hydrocarbons and trace elements in smoke in Merseyside and other Northern localities. *Int. J. Air and Water Poll.* **4**, 141.

50. Surgeon General's Advisory Committee on Smoking and Health : *Smoking and Health,* 1964. Public Health Service Publication. No. 1003 : *a.* pp. 55–58 ; *b.* pp. 150–164 ; *c.* p. 159 ; *d.* p. 182 ; *e.* p. 202 ; *f.* pp. 365–377.

51. Thomas, C. B. 1960. Characteristics of smokers compared with non-smokers in a population of healthy young adults, including observa-

tions on family history, blood pressure, heart rate, body weight, cholesterol and certain psychologic traits. *Ann. intern. Med.* **53**, 697.

52. TOKUHATA, G. K. and LILIENFELD, A. M. 1963. Familial aggregation of lung cancer in humans. *J. nat. Cancer Inst.* **30**, 289.

53. TODD, G. F. and MASON, J. I. 1959. Concordance of smoking habits in monozygotic and dizygotic twins. *Heredity* **13**, 417.

54. WALLER, R. E. 1952. Carcinogens in urban air. *Brit. J. Cancer* **6**, 8.

55. WALLER, R. E. 1965. Carcinoma of bronchus. *Lancet* ii, 953.

56. WALLER, R. E. 1966. The interpretation and use of data on air pollution for epidemiological research. *Statistician* **16**, 45.

57. WALLER, R. E. 1966. Personal communication.

58. WYNDER, E. L. and HAMMOND, E. C. 1962. A study of air pollution carcinogenesis. 1. Analysis of epidemiological evidence. *Cancer* **15**, 79.

59. WYNDER, E. L., LEMON, F. R. and BROSS, I. J. 1959. Cancer and coronary artery disease among Seventh-Day Adventists. *Cancer* **12**, 1016.

60. WYNDER, E. L. and WRIGHT, G. 1957. A study of tobacco carcinogenesis Part I. The primary fractions. *Cancer* **10**, 255.

SOME MAJOR CAUSES OF ILLNESS:
II. PSYCHOLOGICAL ILLNESS
Chairman: Sir Aubrey Lewis

INTRODUCTION

SIR AUBREY LEWIS

Institute of Psychiatry, The Maudsley Hospital, London

THE subject of this session is mental illness. There is no true dividing line between mental illness and physical illness—much mental disorder is the outcome and expression of physical disease—but it is convenient to consider separately those illnesses which chiefly affect thinking, feeling and conduct.

No one could deny that mental illness is a major cause of disability and misery. Consider the number of affected persons, however measured; the distress and the limitations of independent activity that are entailed; the impact upon the patient's family and upon society in general. All these point to its gravity.

It is true that there have been great advances in the treatment of many forms of mental illness, but there is still a substantial burden of unhappiness to be laid at its door. It would be self-deception to suppose that we have now fully controlled or prevented the acute manifestations; the statistics of first admissions to mental hospitals tell a different tale.

In speaking of mental disorder we no longer think only of those gross disturbances which we used to call, collectively, insanity. We include the neurotic illnesses such as hysteria, morbid anxiety and obsessions; disturbances of conduct such as gross sexual perversions; drug addiction; and anomalies of personality.

Inevitably, there has been dispute and uncertainty as to where the line should be drawn between mental disorder on the one hand, and rather extreme types of normal behaviour on the other. But, wherever we draw that line, there is still a large segment of misfortune and pain to be attributed to it, and much need for close study, as well as alleviation by all the means in our power.

The studies reported in this session are all, I think, informed by that spirit. They embody the results of a comparatively recent development in psychiatric research, to reinforce and amplify what we have already learnt from close study of individuals. Our contributors are men who have become acknowledged authorities on the topics listed—prevalence, social influences, alcoholism and suicide. It would be absurd to expect them to deal fully, in the short time at their disposal, with the large and intricate subjects to which they have devoted much research and have acquired much information ; but they are in the best position to judge what are the essentials.

In a pithy introduction to one of his books, Professor Leighton said that he was concerned with three central questions : how much ? of what kinds ? and where ? Expressed in longer sentences, he asks what is the prevalence of mental disorder in the population ? what are the proportions of the different kinds ? and how are the patients distributed in relation to socio-cultural factors ?

These are the issues faced by all the speakers, with varying emphasis, and to a large extent investigated by basically similar methods.

SOME OBSERVATIONS ON THE PREVALENCE OF MENTAL ILLNESS IN CONTRASTING COMMUNITIES *†

ALEXANDER H. LEIGHTON

Department of Behavioral Sciences, Harvard School of Public Health

PRELIMINARY COMMENT

THE field with which this report is concerned—psychiatric epidemiology—is new, and the research work exploratory. Because of this, it is exceedingly difficult to write briefly, and at the same time, be circumspect in statement. To report findings in a way that is truly intelligible, one must first give a detailed description of method; and then, if at the end one ventures into conclusions and interpretations, they should be qualified with great care.

This would require a book rather than a paper, and so of course is not appropriate for this meeting. I compromise, therefore, by attempting a brief statement of some points that appear tenable on the basis of present knowledge, but at the same time making clear that they are little more than suggestions, subject to revision as methods improve and new data emerge. Details of technique and discussion of qualifications are available, as they have been published elsewhere.[6, 7] In this report I give a brief account of the research aims, an outline

* This work has been conducted as part of the Cornell Program in Social Psychiatry and has been supported through funds provided by the Milbank Memorial Fund, the Carnegie Corporation of New York, the Ford Foundation, the National Institute of Mental Health, and the Dominion Provincial Mental Health Grants of Canada. The foundations are not, of course, the authors, owners, publishers, or proprietors of this report, and are not to be understood as approving, by virtue of their grants, any of the statements made or views expressed herein.

† Material in this paper will form part of a book entitled *Determinants of Mental Illness,* edited by Plog, Edgerton and Beckwith, to be published by Holt, Rinehart & Winston, Inc.

Material from the books numbered 6 and 7 in the References at the end of this paper is used with the permission of the publishers of these books.

of the methods employed, a word about the people and areas where the research was conducted and, finally, selected findings together with some inferences that can be derived from them.

RESEARCH AIMS

The original question that concerned my colleagues and myself is this : what role do cultural and social factors have in the production of psychiatric disorders ?

It soon became apparent that this must be broken down into a series of component questions. These include : How much psychiatric disorder is there in populations, regardless of diagnosis and treatment ? What are the proportions of different kinds of disorders ? How are these distributed in relation to cultural and social conditions ? Are there clusterings here and there such as to constitute correlations between disorder and sociocultural factors ?

One realizes, of course, that a correlation does not establish a cause and effect relationship ; but it does set up a target for future examination. This is an expectation common to all forms of epidemiology.

If one makes a prevalence study in one population, the question immediately arises as to how things are in other populations. This leads to drawing samples in various places and comparing them. My colleagues and I have done this a number of times, and I have selected for presentation here two groups in which the work has been the most intensive and extensive.

THE STIRLING COUNTY STUDY [3, 5, 7]

This study was made in a rural and small town county in one of the Atlantic provinces of Canada. The population of 20,000 is about half English and half Acadian French, and is distributed in small communities of several hundred or less, with one town of 3000 inhabitants. A major means used for data collecting was a questionnaire survey administered to a probability sample of the male and female household heads in the county. These data were supplemented by means of interviews with local key informants concerning all the individuals surveyed. In addition, hospital records pertaining to county residents were collected, both locally and from the nearest large metropolitan hospitals.

The psychiatric data obtained concerning each individual in the sample consisted of a review of the systems of the body and a series of questions about psychiatric symptoms. For evaluating these data, we used independent ratings by two or more psychiatrists, who then prepared a joint evaluation of the person stating whether he did or did not show significant psychiatric symptoms, how far his functioning was impaired by such symptoms, and the degree of confidence felt that he was or was not a psychiatric ' case '.

For codification purposes, it was possible to use the terminology of the *Diagnostic and Statistical Manual* (APA, 1952), but we did not try to make diagnoses. Instead, we employed the Manual's terms descriptively and recorded for each person as many different symptom patterns as he showed. With this method, it was possible to achieve a workable degree of agreement between psychiatrists on each of the required judgements.

Meanwhile, a team of social scientists selected communities showing low and high levels of sociocultural integration. This was done initially by interviewing knowledgeable local informants. These communities were then studied intensively by participant observers who usually lived in them. The social scientists also analysed questionnaire survey data according to various sociocultural criteria which included occupational position, education, migration, extent of French or English cultural commitment, and religious involvement.

THE YORUBA STUDY

The research area consisted of about 100 square miles around the Aro Hospital in the western region of Nigeria. This institution for nervous diseases is located near the city of Abeokuta, which had a population of approximately 80,000. The people are almost exclusively of the Yoruba tribe (whose membership numbers about 5,000,000), and Abeokuta is the headquarters for the Egba sub-tribe.

Within the 100 square miles a preliminary selection was made of twenty-five villages, and from these a final selection of fifteen was made, to provide a suitable range in size, in modernization, and in degree of sociocultural integration. Eight segments of Abeokuta were also selected for study.

Psychiatric interviewing was done with a male and a female adult in each household drawn in the villages and towns. The households were drawn according to sampling principles. Anthropological data were gathered concomitantly, again using samples, for the primary purpose of detecting cultural change and of determining various levels of integration. A team of four full-time and two part-time psychiatrists and three social scientists made up the behavioural research group; together with numbers of interpreters and assistants, and a medical team that provided treatment for all comers in the villages while the interviewing was going on. It is to be noted that two of the psychiatrists, Lambo and Asuni, are themselves members of the Yoruba tribe.

Working intensively for three months, we gathered psychiatric data from 262 villagers and 64 residents of Abeokuta. The interviews lasted $1\frac{1}{2}$ to 2 hours each. During the same period, 152 social science questionnaire interviews were obtained from the same sample in the villages and Abeokuta, as well as more free-ranging interviews with headmen and elders.

Evaluations following the Stirling technique were then carried out on the 326 psychiatric interviews.

COMPARISON OF FINDINGS

The distributions of respondents by age and sex for the Nigerian and North American samples are shown in Table I. The use of three age groups rather than a finer breakdown is due chiefly to the difficulty of ascertaining the exact age of the Yoruba respondents.

The similarity of the sex and age distribution in the Yoruba villages to that in Stirling County is striking. We had anticipated that early deaths would reduce the percentage of Yoruba in the middle, and especially the older, age groups. Abeokuta

TABLE I

ALL SURVEY RESPONDENTS BY AGE, SEX, AND PLACE OF STUDY

	AGE GROUP				
	39 or under	40–59	60 and over	TOTAL	
	%	%	%	%	(no.)
Yoruba villages (Nigeria)					
Men	32	40	28	100	(138)
Women	46	28	26	100	(124)
Total	38	35	27	100	(262)
Abeokuta city (Nigeria)					
Men	28	47	25	100	(32)
Women	37	44	19	100	(32)
Total	33	45	22	100	(64)
Stirling County (North America)					
Men	30	40	30	100	(463)
Women	40	36	24	100	(547)
Total	35	38	27	100	(1010)

women differ somewhat from the village pattern, but not significantly.

Table II shows the percentage of A, B, C and D ratings

TABLE II

MAIN PSYCHIATRIC RATING AND IMPAIRMENT
IN SURVEY RESPONDENTS

	YORUBA VILLAGES (Nigeria)	ABEOKUTA CITY (Nigeria)	STIRLING COUNTY (North America)
NUMBER OF RESPONDENTS	262	64	1010
Rating	%	%	%
A	21	31	31
B	19	14	26
A+B	40	45	57
C	35	30	26
D	25	25	17
Total	100	100	100
Significantly impaired	15	19	33

A almost certainly an instance of psychiatric disorder
B probably an instance of psychiatric disorder.
C doubtful.
D almost certainly a psychiatrically well individual.

(likelihood of being a psychiatric case) and the percentage having significant impairment.

The Yoruba villages have the smallest percentage of 'A's (almost certainly psychiatric disorder) and the second smallest percentage of 'B's (probables). If these two are added and compared across the three groups, the result is a steady rise from the villages through Abeokuta to Stirling County. The differences between villages and Stirling in both A and A + B is statistically significant at the 1 per cent level of confidence. The D percentage (or 'well' people) is higher for both groups of Yoruba than for Stirling, a difference that is significant at the 5 per cent level.

The percentage showing impairment also rises in the same direction, with twice as large a proportion in Stirling as in the Nigerian villages (a difference with statistical significance at the 1 per cent level). In both samples, the great bulk of this impairment is mild. The excess in percentage of A + B ratings over the percentage with mild or greater impairment is explainable by the fact that throughout the greater part of his life a person can exhibit psychiatric symptoms without being impaired by them.

The kinds of symptoms found are shown in Table III.

Although the actual percentage of respondents showing symptoms in a major symptom category (psychophysiologic, psychoneurotic, etc.) varies considerably between groups, it is none the less interesting that the overall patterning is similar— there are many psychophysiologic and psychoneurotic symptom patterns and relatively few of the other categories.

Probably more revealing than the direct comparison of percentages between Yoruba and Stirling samples is the comparison of rank order. The major symptom categories of psychophysiologic, psychoneurotic and personality disorder have the same rank order in the villages and in Stirling. Under psychophysiologic, moreover, the villages and Stirling County have the same rank order for gastro-intestinal, musculo-skeletal and cardiovascular. Under psychoneurotic, the 'other' category is much the most common in all three groups, with anxiety or depression coming next. Hypochondriasis was not definable by our methods in the Yoruba sample because of the

TABLE III

CURRENT SYMPTOM PATTERNS AMONG SURVEY RESPONDENTS,
YORUBA VILLAGES, ABEOKUTA AND STIRLING COUNTY

	YORUBA VILLAGES (Nigeria)	ABEOKUTA CITY (Nigeria)	STIRLING COUNTY (North America)
NUMBER OF RESPONDENTS	262	64	1010
Symptom Patterns	%	%	%
Psychophysiologic	81	95	59
Gastro-intestinal	42	50	33
Musculoskeletal	34	52	22
Cardiovascular	20	33	15
Headaches	42	55	12
Respiratory	8	16	3
Genito-urinary	24	47	4
Skin	18	28	1
Endocrine	3	—	2
Overweight	*	—	6
Subjective body sensations	38	41	not used
Psychoneurotic	71	77	52
Anxiety	27	36	10
Depressive	30	27	7
Hypochondriacal	—	—	4
Other	53	61	41
Personality Disorder	7	—	6
Passive-aggressive	3	—	1
Emotionally unstable	2	—	2
Compulsive	—	—	*
Inadequate	*	—	1
Other	3	—	3
Sociopathic behaviour	2	—	6
Alcohol	2	—	3
Dyssocial	—	—	2
Antisocial	1	—	1
Drug Addiction	—	—	*
Mental deficiency	2	—	5
Psychosis	2	—	1
Affective	—	—	*
Schizophrenic	1	—	1
Other	*	—	*
Brain syndrome	5	8	3
Chronic	5	6	2
Convulsive	1	3	*
Other	—	2	1

* More than zero but less than 0·5 per cent.

H

large overlay of organic disorder. The total lack of persons showing symptoms in all four categories in Abeokuta (personality disorder, sociopathic, mental deficiency and psychosis) is probably a function of the small size of the sample.[6]

As to points on which there are major differences in rank order, it can be seen that under ' psychophysiologic ' the Yoruba stand out as apparently suffering from more headaches, gastro-intestinal symptoms, respiratory difficulty, genito-urinary disturbance and skin trouble. Most of this we attribute to the generally higher prevalence of organic disorder, which affects the percentage figures of this whole major symptom category.

In view of the literature on depression in Africa, it is of interest to observe that our percentage figures suggest that the Yoruba have a greater prevalence of depressive symptoms than was found in the North American sample. Not only were these symptoms present in our respondents, but they were common and easily detected in response to the same questions as those employed in Stirling County.

Two categories that are found more frequently in Stirling County are sociopathic behaviour and mental deficiency. It seems reasonable to suppose that there was actually more sociopathic behaviour in the somewhat more complex, less intimate group in North America than in the ' face-to-face ' African village.

In the case of mental deficiency, it seems likely that the apparently lower percentages in Nigeria constitute an artifact. Without the measure of school achievement, this symptom pattern is not easy to detect in its milder forms, and most of the interviewers and evaluators were probably unable to detect indications of mild subnormal mental functioning.

Even though people with sociopathic symptoms are few, it is clear that they consist chiefly of abusers of alcohol, as was the case in Stirling. It seems quite likely that, if we had had more details about the social structure and standards, we might have detected at least some dyssocial individuals.

The somewhat higher frequency of brain syndrome among the Yoruba than in Stirling County may be an indirect effect of the commoner occurrence of organic disease. Because

Table III shows only current symptom patterns and because the sample did not include any persons suffering from an acute brain syndrome, it is worth reporting that the addition of past symptoms would raise the brain syndrome percentage 1·1 per cent for the villages and 4·7 per cent for Abeokuta, but would not change the Stirling County figures at all. On the whole, however, we feel surprised that we did not encounter more evidence of brain syndrome.

SOME GENERAL CONCLUSIONS

1. Despite the many obvious points of social, cultural and biological contrast between the Stirling and Yoruba groups, it would appear that the patterning and distribution of psychiatric disorders are dominated by similarities rather than by differences.

2. It seems, nevertheless, that on the whole the Yorubas do have a lower prevalence rate. One must admit that there are major difficulties in making this comparison. Despite our intention to make the two studies comparable, it is possible that differences in technique and/or in the way people responded to the interview situation could account for the apparently lower figures among the Yoruba. We have given much thought to this; we have re-examined the data, and have done further work in Nigeria. Although there is not at present any way of making absolutely sure, we are inclined to the belief that methodological factors are not sufficient to provide the entire explanation. In other words, our opinion is that in these samples the Yoruba group does have less psychiatric disorder than the Stirling group.

3. In both groups, sex makes a difference to mental health in a striking yet complicated way. Overall, the women of Stirling County show a higher prevalence of psychiatric disorder than do the men. For example, while the percentage of men rated A (almost certainly psychiatric disorder) was 21, the corresponding figure for women was 40 per cent. Conversely, the 20 per cent of the men were rated D (no psychiatric disorder) and 16 per cent of the women. In statistical terms these figures are of course highly significant. Moreover,

the same kind of male-female difference has been found in a number of other groups through studies made in Sweden, Mexico, France and among Eskimos in Alaska.[1, 2, 4, 8] Such findings naturally suggest that there may be biological determinants at work.

The results from the Yoruba study, however, are not in keeping. There the women appear to have the same prevalence of disorder equal to, or lower than that of the men, and a far, far lower rate than the Stirling women. The rank order from high to low of the two sexes across both groups is : Stirling women, Stirling men, Yoruba men, Yoruba women.

The sociological and cultural studies that have been conducted in Stirling County and in the Yoruba villages offer some plausible explanations. In most of the villages the women live according to highly stable (so far) traditions in a world pretty well apart from men and with a minimum of discontinuity between the conditioning experiences of childhood and the fulfilments of adult life. While their world is in many ways limited, it is relatively predictable, providing numerous satisfactions and little in the way of uncertainty, confusion, disorientation, role conflict, and other similar types of stress.

With men, the situation is different. Because of the nature of the male role, particularly in the task of earning a living, they are more exposed to the turmoil, uncertainties and frustrations arising in the massive changes under way in Yoruba society. Even in the most stable of the villages we studied, the men were being affected by these trends and were having difficulties of adjustment, and of knowing what to expect and how to conduct themselves.

In Stirling County (and in many other parts of the Western world, too) a case can be made for concluding that women at the present time are more exposed to stresses associated with role change and role conflict than are men. All people are, of course, more or less exposed to the many stressful consequences of the rapid changes characteristic of the world to-day, but in much of Europe and North America, women have been experiencing in addition a special set of adaptations involving work, career, family, and shifting conflictful criteria of prestige.

These considerations led us to re-examine the Stirling data.

One community was identified in which the mental health of women appeared better than that of the men. It was found also that in that community a strong traditional culture gave women a protected position and many rewards for performing well as wife, mother and grandmother, whereas the men were more exposed to adjustment difficulties emanating from the larger society.

In another Stirling community, the sociocultural environment was more favourable to men than to women. In it the men had a very low prevalence of psychiatric disorder, while the women had a rate somewhat higher than the County average. Finally, in certain communities where severe and numerous stress factors affected both sexes, men and women had very high, but approximately equal, prevalence rates.

From all this it would seem that both men and women are affected by social and cultural circumstances—that there is something more at work than biologically determined characteristics. It seems equally evident, however, that at given times and in a given place men and women may be affected very differently; that what is a benign environment for one sex may be stressful and noxious for the other. We have, then, a matter worth further inquiry, since it can obviously have profound effects on family formation, child rearing and intra-group relations—indeed on the whole fabric of society.

4. The degree of sociocultural disintegration in the community in which people live makes a difference to their mental health.

Our prototypes of disintegration are certain small neighbourhoods in Stirling County. A disintegrated group is one that lacks leadership, the capacity to follow leadership, and the ability to make collective decisions; it also displays poor intra-group communication and weak and conflicting values regarding conduct; economic resources are generally low. The word ' demoralized ' may perhaps summarize their condition. Sociological and anthropological techniques have been developed for identifying and roughly measuring degrees of disintegration, and this can be done for the most part independently of the assessment of psychiatric disorder prevalence.

One of the most clear-cut findings in the Stirling County

Study is that there is a very high correlation between socio-cultural disintegration and the prevalence of psychiatric disorder. A tenable hypothesis is that this is mainly due to heredity, and that these disintegrated neighbourhoods are the product of years of indrifting and inbreeding of inadequate people. It can be supposed that disintegrated communities are being produced by disintegrated personalities. The argument is harder to support in the case of the Nigerian villages; yet the same kind of correlation between disorder prevalence and sociocultural disintegration appears in them also. In this case, however, the disintegration is recent and has come upon the villages through extraneous forces.

It is my conclusion that the evidence so far tends to favour—although it does not prove—the hypothesis that disorganization of social systems is in and of itself likely to generate psychiatric disorders. We may suppose it does so through damage to the personalities of children during the formative years, and also through the effects of contemporary stress upon the adults.

In some ways, findings like these are frightening. We live in a world in which social systems are falling apart on all sides. There is, of course, a counter-trend of rebuilding the old and establishing new systems, but the disorganization is notable. Our findings seem to indicate that such a process tends to create psychological states that increase the disorganization and so condemn us to descend in ever increasing vicious spirals.

More optimistic is the thought that the preventive measures envisaged are in harmony with the counter-trend of numerous programmes designed for improving education, health, welfare, recreation and opportunity for work. But such optimism is tinged with a sense of urgency because of the need to make progress before the spirals become too widespread, too deep, and too difficult to reverse.

DISCUSSION

Asked whether the correlations between age and psychiatric disorder showed great differences as between North America and Africa, PROFESSOR LEIGHTON said that, in his paper, he had had to bypass the age distribution. There was a very big age factor; younger people tended to have far fewer psychiatric disorders than older people. As people got older the frequency of disorder increased in both sexes, though in women it increased faster than in men. There then tended to be a levelling off in middle life—at least in the North American sample— a further rise round the retiring age, and then a decline. In the generalizations he had made in his paper, both age and sex factors were controlled.

In reply to another questioner Professor Leighton said that the members of the survey team had been aware of the problems surrounding the interviewing of Nigerian villagers, and had done their best to cope with them. For about two years before the survey started their Nigerian colleague, Dr Lambo, had had one of his men going around the villages to discuss the coming of a health survey team. Talks had been held with the Elders to promote the idea that this would be a good plan; the man who did this contact work had appropriate kinship ties in the villages. Dr Lambo also obtained the approval of the Paramount Chief of the region, and word was passed down through the regional chiefs.

During the survey a medical clinic had provided examination and treatment to all the people in the village who wished to take advantage of its services. It had seemed probable that the presence of a medical team would make the villagers more, rather than less, likely to report symptoms.

With regard to language, the American and Canadian members of the research group had spent a year studying the Yoruba language—not expecting to speak it fluently, but hoping to be able to understand the problems of interpretation; a Yoruba speaker had helped to draw up the questionnaire. Dr Lambo had had a number of medical students and psychiatric nurses working with him who were themselves members of the Yoruba tribe, and there two Yoruba re-translations had been made of the questionnaire. It had been tried out in a village which was not part of the sample, and had been again revised; it had also been tested on sixty known psychiatric patients. During the survey the psychiatric interviewer had been accompanied by a Yoruba interpreter who had done the actual asking; the interviewer had been, in most cases, either a medical student trained over a number of months for this work, or a psychiatric nurse, or someone with similar qualifications.

The attitude of the Yoruba towards mental illness, Professor Leighton said, was apparently much like that in Europe and America: mental illness was not a good thing to have in the family; it could interfere with a desirable marriage. Nevertheless, when people did become ill, they were treated with a good deal of consideration and kindness.

In answer to a question about the influence of religion, Professor Leighton said that there were two points to consider: first, did religion seem to have a good effect in preventing mental illness; and second, did it play a part in helping back to health those who were mentally ill? In the Stirling County study, about half the population had been Roman Catholics; the remainder were divided up between several different Protestant denominations, Baptists being the most numerous, and followers of the United Church of Canada second. No significant correlation had been found between mental health and any particular religious group, or between people who disavowed religious connections and those who had some religious leanings. On the other hand, the church was an important part of the organization of most communities, and the communities which had good integration had the best mental health; so the church was evidently one of the factors making for mental stability.

Asked whether there was any difference between the two groups, or between the sexes, in the incidence of psychotic illness, Professor Leighton said that in their samples—with perhaps 1 per cent psychotic—the figures were too small for a useful analysis.

The question of the belief in witchcraft in Nigeria was raised. The survey team had made a study of native practices and had, in fact, examined twelve patients who were being treated by native doctors. In Nigeria witchcraft was considered to be one of the commonest causes of mental illness, but that did not contradict the concept of *illness*. The patient was still regarded as mentally deranged even if witchcraft was believed to have been the cause of his illness.

REFERENCES

1. BRUNETTI, P. M. 1964. A prevalence survey of mental disorders in a rural commune in Vaucluse; methodological considerations. *Acta psychiat. scand.* **40**, 323.
2. ESSEN-MOLLER, ERIK. 1956. Individual traits and morbidity in a Swedish rural population. *Acta psychiat. neurol. scand.* Supplement 100.
3. HUGHES, C. C., TREMBLAY, M., RAPOPORT, R. N. and LEIGHTON, A. H. 1960. *People of Cove and Woodlot.* Communities from the viewpoint of social psychiatry: Vol. II, The Stirling County Study of Psychiatric Disorder and Sociocultural Environment. New York. Basic Books.
4. LANGNER, T. S. 1965. Psychophysiological symptoms and the status of women in two Mexican communities. In *Approaches to Cross-cultural Psychiatry* (Ed. J. M. Murphy and A. H. Leighton) pp. 360–392. Ithaca, N.Y. Cornell University Press.
5. LEIGHTON, A. H. 1959. *My Name is Legion.* Foundations for a theory of man in relation to culture: Vol. I, The Stirling County Study of Psychiatric Disorder and Sociocultural Environment. New York. Basic Books.
6. LEIGHTON, A. H., LAMBO, T. A., HUGHES, C. C., LEIGHTON, D. C., MURPHY, J. M. and MACKLIN, D. B. 1963. *Psychiatric Disorder among the Yoruba:* A report from the Cornell-Aro Mental Health Research Project in the Western Region, Nigeria. Ithaca, N.Y. Cornell University Press.
7. LEIGHTON, D. C., HARDING, J. S., MACKLIN, D. B., MACMILLAN, A. M. and LEIGHTON, A. H. 1963. *The Character of Danger.* Psychiatric symptoms in selected communities: Vol. III, The Stirling County Study of Psychiatric Disorder and Sociocultural Environment. New York. Basic Books.
8. MURPHY, JANE M. 1962. Cross-cultural studies of the prevalence of psychiatric disorders. *Wld. ment. Hlth.* **14**, No. 2.

SOCIAL INFLUENCES
ON THE PREVALENCE OF
PSYCHIATRIC DISORDER

KENNETH RAWNSLEY

Department of Psychological Medicine, Welsh National School of Medicine, Cardiff

ALTHOUGH the title of my paper refers to psychiatric disorders, I shall have occasion to speak of somatic ailments too, since the social factors I wish to discuss may influence the prevalence of both categories of illness in similar fashion.

The Ministry of Pensions and National Insurance published recently [7] the findings of an inquiry into the incidence of incapacity for work, covering the whole of Great Britain. The survey was made during a twelve-month period beginning June 1961. The data were obtained from certificates presented to the Ministry in support of claims for sickness benefit. Special attention was paid to certain medical conditions, namely bronchitis, arthritis and rheumatism, psychosis and psychoneurosis. Rates for the inception of disability were calculated for standard regions of England and also for Wales and Scotland. There was substantial variation in the rate for incapacity due to psychosis and psycho-neurosis, by area of the country (Table I). There was also considerable variation

TABLE I

MALES COMMENCING ONE OR MORE SPELLS OF INCAPACITY FOR WORK
DUE TO PSYCHOSIS AND PSYCHONEUROSIS IN SURVEY PERIOD,
PER 1000 MEN

REGION	RATE PER 1000	REGION	RATE PER 1000
Wales	12·1	South Western	6·9
North Western	9·0	Midland	6·7
North Midland	8·1	London and South Eastern	6·1
East and West Ridings	8·0	Eastern	5·2
Northern	7·5	Southern	5·2
Scotland	7·3		

in the rate due to all causes of incapacity (Table II). This latter rate is based on individual persons rather than on events ; in other words, if a person was absent from work on one

TABLE II

MALES COMMENCING ONE OR MORE SPELLS OF INCAPACITY FOR WORK
FROM ALL CAUSES IN SURVEY PERIOD, PER 1000 MEN

REGION	RATE PER 1000	REGION	RATE PER 1000
Wales	362·6	South Western	263·0
Northern	331·4	Midland	257·2
North Western	323·3	London and South Eastern	250·1
East and West Ridings	307·9	Eastern	239·6
Scotland	293·5	Southern	229·3
North Midland	271·7		

occasion due to a particular cause, or on several separate occasions throughout the year due to different causes, this was counted as one instance of incapacity.

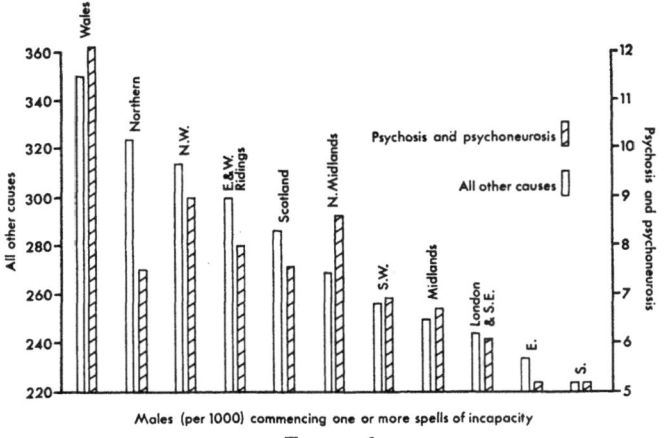

Males (per 1000) commencing one or more spells of incapacity

FIGURE 1

Males (per 1000) commencing one or more spells of incapacity due to psychosis or psychoneurosis and (separately) due to all other causes, by region of Great Britain.

A comparison of the rate of incapacity due to psychiatric causes with the rates for all other causes of morbidity in different areas of the country shows a fairly close correspondence in the rank order of these two indices (Figure I).

Data of this kind are of course very crude reflections of illness as found in the general population. Incapacity for work is only one facet of illness. Rate variations may indicate differences in willingness to seek advice from doctors and/or in readiness to take time off work for sickness of a given severity.

The variations between regions of the country are certainly rather striking, however, and, in particular, the co-variation of rates for psychosis and psycho-neurosis with the rates for all other causes of morbidity invites comment. Many of the patients with psychiatric diagnoses may also have lost time at work during other parts of the survey year for other reasons. It is not possible to tell from the Ministry data to what extent the same persons may have had multiple diagnoses during the year.

The concordance of rates for psychiatric and of rates for other causes of morbidity between large sections of the population prompts the question : to what extent is there an association between psychological ailments and somatic disorders in smaller units of population, e.g. in families, and also in individuals ?

In a study by a psychiatrist, carried out in general practice, members of families of psychiatric patients showed an unusually high prevalence both of psychiatric and of physical illness.[1] Buck and Laughton [2] examined the medical records of sixty-five families. Children in families where the mother had a history of psychiatric illness had a comparatively high proportion of physical and of psychological disorder.

As regards individuals, Downes and Simon [4] in the course of a longitudinal study of 828 families singled out ninety in which one or more neurotics were present. These individuals were found to have more than average chronic illness, and acute physical illness, and accidents. Doust [3] compared 272 psychiatric patients with various diagnoses with 354 controls. The patients showed a comparatively high incidence of physical illness. Longaker and Godden [6] found that in a random sample from a small town in Eastern Canada, individuals impaired because of psychiatric symptoms were more likely than others to give a history of organic illness. Brown [1] found in the course of a survey in general practice that patients presenting a psychiatric disorder showed a substantially higher prevalence of physical illness during the course of a year than those who had no psychiatric disorder.

I should like to present some evidence from work carried out in the Medical Research Council Social Psychiatry Research Unit in South Wales to investigate the association between psychological and somatic symptoms and disorders. A survey of the prevalence of a wide range of symptoms, together with an inquiry into attitudes to these symptoms, was made in a rural population in South Wales where a private census with social information about each individual was available. A stratified random sample of the population of this area aged 26–45 was surveyed through the medium of household interviews using closely structured procedures, through information taken from general practitioners, and through collection of material from psychiatric clinics and hospitals. The stratification was based on an *ad hoc* method of social classification of the entire population into six categories; the details are not relevant here. In the household interview, one of the methods used for collecting information about symptoms was a modification of the Cornell Medical Index Health Questionnaire.[8]

The Cornell symptoms may be divided into those which are ostensibly somatic in nature, such as pain in the chest, cough, or swelling of the ankles, and those which are evidently psychological in type, such as tension or morbid fears. It is possible to examine the association between the frequency of somatic symptoms in individuals with the frequency of psychological symptoms in the same individuals. For the random sample of the South Wales rural population the correlation coefficients are positive (Table III).

Further evidence about the association between physical and psychological disorders is derived from work carried out by the Medical Research Council Unit among the population of Tristan da Cunha, after their evacuation to England following the eruption of a volcano.[9] This work formed part of a large-scale investigation into the health of this community of about 260 individuals, made by the Medical Research Council.

By good fortune, the records of a Norwegian expedition to

TABLE III

RURAL POPULATION SURVEY: ASSOCIATION BETWEEN FREQUENCY
OF PHYSICAL AND OF PSYCHOLOGICAL SYMPTOMS IN INDIVIDUALS
BY SOCIAL SECTION

| SOCIAL SECTION | Product moment correlation coefficients | |
	MALE	FEMALE
A	+ 0·35	+ 0·60
B	+ 0·44	+ 0·66
C	+ 0·41	+ 0·63
D	+ 0·48	+ 0·47
E	+ 0·59	+ 0·61
F	+ 0·66	+ 0·68

the island, made in 1937, and including reports by medical men and by a sociologist, were available. The arrival of the expedition coincided, by chance, with the outbreak of an epidemic of hysterical behaviour among a substantial minority of the island population, mainly among the younger women. The records were detailed and comprehensive, and it was possible to identify by name those members of the population who had manifested hysterical behaviour in 1937. The majority of these individuals survived and came to England in 1961.

In the course of our survey of the community, Dr Loudon and I were quickly struck by the frequency with which headache was reported as a symptom. The headache was frontal and was described in characteristic, almost stereotyped, words and gestures. In many cases anxiety and worry were mentioned as provoking causes. Other precipitants were bright sunshine, strong winds and the menses. These headaches were endemic both on Tristan and in this country.

For purposes of description we may call headaches with a definite history of emotional provocation, psychogenic (Table IV). The frequency of psychogenic headache among those islanders known to have had attacks of major hysteria in 1937 was significantly higher than the frequency in a control group of islanders matched for sex and age (Table V).

A further source of information about morbidity was the record of consultations with the island doctor, Mr Norman

TABLE IV

PREVALENCE OF HEADACHE

	MALE		FEMALE		BOTH SEXES	
	No.	%	No.	%	No.	%
Psychogenic headache	13	14 ⎱ 52	32	34 ⎱ 66	45	24 ⎱ 59
Non-psychogenic headache	36	38 ⎰	30	32 ⎰	66	35 ⎰
No headache	45	48	31	33	76	41

TABLE V

PREVALENCE OF PSYCHOGENIC HEADACHES IN THE
1937 CASES AND IN CONTROLS

	1937 CASES	CONTROLS
Psychogenic headache	16	3
Remainder	3	16

$$\chi^2 = 15 \cdot 2 \qquad p < 0 \cdot 001$$

Samuels, who continued to look after the community for a period after arrival in England.

The frequency of consultations, for whatever reason, of the group of islanders with previous history of hysteria has been compared with the frequency for a matched control group for a seven-month period from June 1961 to January 1962. The 1937 cases have a notably higher mean consultation rate than have the controls (Table VI).

TABLE VI

FREQUENCY OF MEDICAL CONSULTATIONS
DURING PERIOD 12.6.61–22.1.62

MEAN NUMBER OF CONSULTATIONS
1937 cases	(19)	7·8
Controls	(19)	4·4

$$t = 2 \cdot 2 \qquad p < 0 \cdot 05$$

Twenty-six specified varieties of illness or of symptom were recorded for the two groups together, irrespective of the number of consultations per disorder or symptom. In nineteen of them a greater proportion of the 1937 cases than of the controls were represented. These conditions included cough, headache, vomiting and nausea, diarrhoea, infective hepatitis, injuries, burns and strains. The control group preponderated in only four disorders and equal proportions were found in three disorders. To make the same comparison in a slightly different fashion, a total of 145 person/complaints were recorded for both groups—92 for the 1937 cases and 53 for the controls.

Somewhat similar findings emerge from a comparison of the consultation behaviour of the islanders who, in 1962, were subject to psychogenic headaches, with that of a matched control group. The headache cases showed a higher rate for all consultations (Table VII).

TABLE VII

FREQUENCY OF MEDICAL CONSULTATIONS
DURING PERIOD 12.6.61—22.1.62

MEAN NUMBER OF CONSULTATIONS (both sexes)

| Psychogenic headache (45) | 7·2 |
| Control group (45) | 4·3 |

$$t = 2·7 \qquad p < 0·01$$

Disregarding the number of consultations per disorder or symptom, thirty-two specified varieties of illness were presented by the ninety individuals comprising the psychogenic headache group and the control group. In twenty-three disorders, which included cough, vomiting and nausea, epigastric pain, infective hepatitis and septic lesions, a greater proportion of the psychogenic headache group than the controls were represented. The control group preponderated in only three disorders, and equal proportions were found in six conditions. Put in a different way, out of 313 person/complaints, 189 were accounted for by the psychogenic headache group and 124 by the controls.

How are we to interpret these findings? It is possible that those islanders who have shown themselves in the past to be susceptible to hysteria are also prone to develop a wide variety of complaints including infective hepatitis, cough and so on.

Alternatively, this group of islanders may be no more prone to develop such ailments than the rest of the population. It may simply be that illness, whatever its nature, is more likely to be brought to the attention of doctors by the members of this group than by their fellows.

Many reasons may be advanced to account for this being so :

1. The 1937 cases, by virtue of their disposition, may be prey to a number of psychogenic complaints which bring them to the doctor frequently. Accordingly they are at greater risk of having other disorders detected and noted, disorders which may or may not have a psychogenic component.

2. The 1937 cases, again for constitutional reasons, may be more readily disabled or discomfited by a given illness than are the run of islanders and may thus be more likely to seek professional aid.

3. Associated traits of personality in the 1937 cases may operate to bring the individual into regular contact with the doctor. I am thinking now of a tendency to morbid preoccupation with health matters or even of a basic sense of insecurity leading to an inclination to seek reassurance and support from someone who can speak with authority.

To recapitulate, it seems therefore that according to the Ministry of Pensions material the frequency of psychiatric disorders and the frequency of somatic ailments vary in a concordant fashion between different regions of Britain. This co-variation may also be discovered among members of families, and also within individuals. Let us consider the functional relationship between these two categories of symptom or disorder in the individual. It may be that both categories when they are present together are reflecting a common pathology. On the other hand, the somatic disorder may give rise to psychological disturbances or vice versa. There is some evidence that both psychological and somatic disorders occur together in crops over time in relation to stressful circumstances.[5] Another factor which must be borne in mind, however, is the possibility that some individuals or groups of individuals may have a relatively low threshold for declaring the presence of symptoms which they experience to GPs and

I

to survey interviewers.   This may apply to both psychological
and somatic disorders.

Among families, the association between ill-health in spouses
may be based upon assortative mating, or may arise by direct
influence of one member upon the other.   Besides, the question
of differential threshold for declaring ills must also be remembered in this connection.

The positive association between physical and psychiatric
disorders in large sections of the population, as is apparent
from the Ministry of Pensions survey, may depend upon many
factors.   Obviously, if psychological disorders may give rise
to physical ailments, or vice versa, then in those sections of
the population where for any reason there was a high prevalence of one or other category of disorders, the other might
be expected to follow suit.   Following the work of Hinkle,[5]
however, one might postulate that stressors, present in certain
sections of the population or in certain areas of the country,
might give rise to both categories of disorder in a large number
of individuals at one time or another.   Again, however, atti-
tudinal factors bearing upon the declaration of illness, readiness
to seek advice, willingness to stay off work, depending perhaps
upon both economic conditions and sets of attitudes and values
prevailing in the areas, might account for the apparent concordance between incidence of somatic and psychological
disorders.

According to the Ministry of Pensions inquiry, Wales had
the highest rates in Great Britain for incapacity for work due
to psychiatric causes and also the highest rates due to all other
causes taken together.   Within Wales certain areas in the
South Wales coal mining valleys show extremely high rates
for both psychiatric and other causes of incapacity.   Preliminary
results from a survey at present being carried out by members
of an MRC team indicate that the population of these same
areas in the mining valleys have substantially higher rates of
admission to psychiatric in-patient and out-patient services
than have the population of an adjacent rural area.   The
present survey will include examination by home interview of
a random sample of the valley population.   The methods
employed for the assessment of psychological and somatic

symptoms and of attitudes to symptoms and to medical care are precisely similar to the methods used in surveying the nearby rural population.

It will be interesting to see whether the higher prevalence of hospital-treated psychiatric illness in the mining valley is paralleled by a higher prevalence of symptoms among the random sample of population; and also, whether this difference, if found, is associated with differences in attitudes.

The social influences which may contribute to the genesis of psychiatric disorder are manifold. I have mentioned those attitudes and values which influence the way in which we experience, evaluate and declare our symptoms. Although partly the product of individual personality these are, to a degree, socially determined, and to that extent are perhaps susceptible of modification by social measures.

REFERENCES

1. BROWN, A. C. 1965. The general morbidity of neurotic patients. MD thesis. University of Cambridge (unpublished).
2. BUCK, C. W. and LAUGHTON, K. B. 1959. Family patterns of illness: the effects of psychoneurosis in the parent upon illness in the child. *Acta psychiat. neurol. scand.* **34**, 165.
3. DOUST, J. W. L. 1952. Psychiatric aspects of somatic immunity: differential incidence of physical disease in the histories of psychiatric patients. *Brit. J. prev. soc. Med.* **6**, 49.
4. DOWNES, J. and SIMON, K. 1953. The characteristics of psychoneurotic patients and their families as revealed in a general morbidity survey. *Psychosom. Med.* **15**, 463.
5. HINKLE, L. E. 1961. Ecological observations of the relation of physical illness, mental illness and social environment. *Psychosom. Med.* **23**, 289.
6. LONGAKER, W. D. and GODDEN, J. O. 1960. A comparison of organic and psychiatric symptoms in a small town. *Acta psychiat. neurol. scand.* **35**, 91.
7. MINISTRY OF PENSIONS AND NATIONAL INSURANCE. 1965. *Report on an Enquiry into the Incidence of Incapacity for Work.* Part II. Incidence of Incapacity for Work in different Areas and Occupations. London. HMSO.
8. RAWNSLEY, K. 1966. Congruence of independent measures of psychiatric morbidity. *J. psychosom. Res.* **10**, 84.
9. RAWNSLEY, K. and LOUDON, J. B. 1964. Epidemiology of mental disorder in a closed community. *Brit. J. Psychiat.* **110**, 830.

GENETIC AND SOCIAL INFLUENCES
IN THE CAUSATION OF SUICIDE

ERWIN STENGEL

Department of Psychiatry, University of Sheffield

IT is generally taken for granted that genetic as well as social factors play a part in all normal and abnormal behaviour. There are some patterns of behaviour for which the co-existence of both these factors has so far not been clearly demonstrated. Suicide is such a pattern. The role of social influences in its causation is well documented. They have, in fact, been more frequently and more thoroughly investigated than any others, owing to the early involvement of sociologists in suicide research. The literature about the psychopathology of suicide is less extensive, and very little has been written about the role of genetic factors. Some clinical and epidemiological observations have tended to suggest that heredity does play a part in the causation of suicide. Familial incidence of suicide has been noted by many workers, but the published data are not comparable and there is a lack of controlled studies. For instance, Dahlgren [1] found that in 6 per cent of 237 cases of attempted suicide admitted to hospital at Malmö, Sweden, there had already been a suicide in the family; but his sample was unrepresentative. The same applies to two of my samples of attempted suicides [5] in which the familial incidence of suicide had been 8 and 10 per cent respectively. For obvious reasons, this kind of information is even more difficult to obtain in cases of suicide than in suicidal attempts, where the key informant is available for questioning. Sainsbury, [4] studying coroners' records, found a family history of suicide in 2·6 per cent of 390 cases of suicide in which a mental disorder or abnormal personality had been a principal factor. Unreliable though the figures are, they indicate that the incidence of suicides in the families of people who had committed suicidal acts is exceptionally high. These observations have been sufficiently common to make a family history of suicidal acts an indicator of a heightened suicidal risk in individual cases. Familial incidence is, of course, no proof of genetic influences in the causation of a phenomenon, but it does invite the hypothesis that such influences are at work. I am aware of only one attempt at testing this hypothesis, viz. the study of Kallmann and Anastasio [3] published in 1947. As that study has received very little attention, I should like to review it here. It was carried out in the Department of Medical Genetics of the New York State Psychiatric Institute.

The authors referred to the report on the accumulation of suicides in certain families and mentioned Kallman's observation that the suicide rate among the blood relatives of schizophrenics consistently exceeded that of a comparable general population, the increase being more pronounced for the patient's children and siblings than for their nephews and nieces. The authors did not expect a specific combination of genetic factors determining the occurrence of suicides as such, because there was no inheritance of any finished human traits, but only a transmission of potential capacities for the development of these traits under certain life conditions. Genetic elements may, however, play a significant part either in the formation of personality types liable to react to stress by suicide, or in the development of special forms of mental disease which favour the tendency to self-destruction. Study of twins appeared to be the only experimental method by which to separate environmental from biological factors in the causation of suicide. In reviewing the psychiatric literature the authors could find only six references to twin suicide cases, only two of them from the last fifty years; they had been twin pairs suffering from manic-depressive psychosis.

A review of their own material of about 2500 twin index cases supplied a total of eleven twin pairs with suicide of one member only. Their search for twin pairs with suicide of both partners had been as unsuccessful as that of other investigators who had pursued psychiatric twin studies. Of Kallman and Anastasio's eleven twins who had committed suicide, three were classed as monozygotic. The life-span available to the surviving twins from the time the co-twins took their lives amounted to an average of seventeen years. The family histories of the eight binovular twins showed a considerable number of psychoses in the close consanguinity, with a predominance of endogenous depressions and a total of seven additional suicides. The life histories of those eight non-identical twin pairs failed to throw light on what had made one of the twins commit suicide, considering that the majority of them had shared the same environment over much of their lives. The authors remarked that in two cases it was difficult to understand why it had not been the other twin who had found life unbearable.

The life histories of the three pairs of identical twins were reported in some detail. The first pair were sisters, one of whom killed her husband and herself to evade deportation to a concentration camp. Her twin sister died at the age of fifty-one from a cerebral haemorrhage on receiving the news of the suicide. The second pair were aged sixty-one when one of them killed himself in a state of depression. There were indications of intense sibling rivalry, but otherwise the similarities in their life histories far outweighed the dissimilarities. The environmental and psychiatric backgrounds of the third pair of identical twins were significantly different. One of the twins had a congenital heart lesion, which resulted in differences of upbringing. At the age of sixteen she developed an acute schizophrenic condition, which subsided after eight months. Her twin sister developed a similar, though less severe, condition three years later. The first mentioned twin killed herself while receiving treatment for a syphilitic infection, the other had a short relapse of her schizophrenia, but afterwards she continued at home without serious difficulties. Comparing the histories of the twins who ended by suicide with those of the surviving co-twins the authors found differences which had been entirely in favour of the survivors.

They concluded that suicide did not occur in both members of twin pairs, even if they were alike in type of personality, cultural setting, social frustration and depressive features of a psychosis. They felt justified in drawing this conclusion from their small material because they believed that, if concordance of twin partners in the matter of suicide had been observed, this finding would have been published. They offered two possible explanations for the apparently consistent discordance of twin pairs as to suicide. The one would incriminate sibling rivalry; but they argued, not very convincingly, that if this was the cause one would expect people who had no siblings to have a much lower suicide rate than people with siblings. There was no evidence that this was so. They therefore favoured another explanation, namely "that suicide seems to be the result of such a complex combination of motivational factors as to render a duplication of this unusual constellation very unlikely, even in identical twin partners who show the same type of psychosis and a very similar degree of social privation." This explanation does not, of course, rule out sibling rivalry as one of the causes of the discordance.

Kallmann and Anastasio's study can be criticized on two grounds. The identity of their supposedly monozygotic twins was not established beyond doubt and the life histories of their patients were somewhat incomplete by modern standards. Their conclusions are nevertheless highly suggestive. The most remarkable, and rather surprising, aspect of this work in the context of this symposium, is the conclusion that known genetic influences in the aetiology of a mental disorder, such as depressive illness, cannot by themselves account for the occurrence of suicides among blood relations. The hypothesis, then, that the familial incidence of suicide is sometimes due to an underlying mental disorder in whose aetiology heredity plays a part, seems to have been refuted by one of the leading experts in the genetics of mental illness. At any rate, he and his associate failed to demonstrate genetic influences in the causation of suicidal acts and passed the whole problem over to the environmentalists. Their findings do not, of course, imply that genetics plays no part in the causation of suicide but that it can play a part only under certain environmental conditions.

An investigation by Walton[7] throws some light on the nature of those conditions. In a previous study, Walton had confirmed the finding of other workers that a history of a broken home in childhood increased the likelihood of suicidal acts in adult life. He now set out to test the following hypothesis. Patients suffering from depressive illness who do and those who do not commit suicidal acts in the course of their abnormal depression differ with regard to the history of parental deprivation.

The method employed in this investigation was as follows. The case notes of all patients with depressive illness admitted to the Maudsley and the Bethlem Royal Hospital during 1955 were examined. The number of cases so diagnosed was 223. They were separated into two groups : (a) the non-suicidal depressives, i.e. those who were not recorded as having attempted or explicitly threatened suicide—they numbered 163 ; (b) those who had—they numbered sixty.

The two groups were examined for the incidence of three factors, each of which has been found to be correlated with the suicide rates in the general population : (1) *Parental deprivation* (' broken home ') which was defined as loss of a parent before the age of fourteen, violent discord among the parents and prolonged estrangement from one of them ; (2) *Social isolation*, the patient being classified as ' socially isolated ' if at the time of becoming ill he lived alone in private rooms or a boarding house or an hotel ; (3) *Manifestation of ' social degeneration '*, i.e. divorce, illegitimacy, loss of a job.

Statistical analysis showed a significant association between membership of the suicidal group and the occurrence of parental deprivation during childhood. No significant association of suicidal behaviour with the other two factors could be demonstrated in this sample.

Walton's finding is of considerable theoretical and practical importance. It means that ' when a patient with depressive illness behaves suicidally, other factors than those producing the depressive mood must have operated '. Kallmann and Anastasio came to the same conclusion. Thus, a history of parental deprivation in childhood is relevant in the clinical assessment of the degree of suicidal risk in individual cases of depressive illness.

We have to be cautious, however, in applying experimental findings, such as those of Walton, to clinical practice before they have been confirmed in representative samples. Walton's failure to confirm the correlation between the other two social factors and suicidal behaviour is open to various interpretations. He did not regard his sample as representative. Possibly in depressive illness those factors do not play the same part in the origin of suicide-proneness as they do in the general population, because the depression itself isolates and divorces the person from society.

Walton's study illustrates some of the difficulties suicide research is up against. He correlated clinical with sociological data. The latter are generally held to be ' hard ' data representing facts, compared with the ' soft ' psychological data which are based on hearsay, i.e. the patient's statements, and clinical impressions only. But how ' hard ' are the sociological data ? Walton's definition of the ' broken home ' or ' parental deprivation ' is more precise than that of other workers, but it is still far from purely factual. Once one goes beyond taking account of age, sex, number of siblings, etc., one is no longer dealing with objective data. It is often difficult to decide what kind and what degree of parental discord qualifies for a home to be retrospectively classed as ' broken '. Yet if we included only families where a parent had been absent or lost we should be rightly criticized for ignoring important psychological and social factors. The definition of ' social isolation ' does not seem to present similar difficulties. What could be more precise and objective than the definition adopted by Walton ? One needs only to look up the last Census to find out how many people in a certain area lived in private rooms, boarding houses or hotels to establish the rate of social isolation. But this information will satisfy only those workers who confine themselves to the analysis of statistical data and avoid talking to people. Once one breaks this taboo, which many sociologists never do, one runs into trouble with the traditional concept of social isolation. This has happened to Peter Townsend,[6] now Professor of Sociology in the University of Essex, who studied the family life of old people in the East end of London. He found the conventional concept of social isolation uninformative about social relations. Many old people, who by definition fell into this category, maintained close personal contacts with their families. Townsend measured the degree to which a person maintained social relationships by counting the frequency of personal contacts within a certain time. He proposed a new concept, that of ' social desolation ', for real deprivation of relationships, due to death, emigration, etc. All this goes to show that some of our supposedly most reliable instruments used for the measuring of social relations and the degree of social integration are still pretty crude.

Even the suicide rates, which have often been described as the most reliable epidemiological data available, are not above suspicion. An investigation into the methods of ascertainment of suicide in different countries, which is still in progress, has cast serious doubts on the comparability of suicide rates obtained by different methods of ascertainment. For instance, suicide rates based on coroners' verdicts are not comparable with rates derived from police surgeons' reports. This means that the suicide rates of the United States and this country cannot be compared with those of Sweden and Denmark. It is doubtful whether even the English and Scottish suicide rates are comparable, because they are not arrived at by identical methods of registration. It seems, then, that many of the supposedly hard data with which suicide research has been working are sadly lacking in consistency. Since Durkheim, many sociologists have maintained that psychological factors play no part in the suicide rates. Gibbs and Martin,[2] two American sociologists who recently attempted to develop an instrument for the measurement of social integration, assess psychological morbidity by the number of mental hospital beds. To psychiatrists this is unacceptable. The antithesis between social and psychological factors in the causation of suicidal behaviour is an artefact created by the differences in methods successively employed in suicide research. Social and other factors, such as age, sex and illness, influence behaviour and may thus increase or inhibit suicide proneness. To do so they have to become psychological factors which, however, are at present more difficult to define and to measure than social factors. On the other hand, psychological changes, such as depressive illnesses, tend to transform the individual's social situation. The separation of social from psychological factors in the causation of suicidal behaviour is no more than a methodological convenience or even necessity. One becomes aware of this if one tries to explain the familial incidence of suicide referred to earlier in this talk. Suicide research is the concern of workers in many fields, and no single method or approach can claim to be the most important.

REFERENCES

1. DAHLGREN, K. G. 1945. *On Suicide and Attempted Suicide.* Lund. Lindstedts.
2. GIBBS, J. P. and MARTIN, W. T. 1964. *Status Integration and Suicide.* Oregon. University of Oregon Books.
3. KALLMAN, F. J. and ANASTASIO, M. M. 1947. Twin studies on the psychopathology of suicide. *J. nerv. ment. Dis.* 105, 40.
4. SAINSBURY, P. 1955. *Suicide in London.* London. Chapman & Hall.
5. STENGEL, E. and COOK, N. G. 1958. *Attempted Suicide.* London. Oxford University Press.
6. TOWNSEND, P. 1957. *The Family Life of Old People.* London. Routledge & Kegan Paul.
7. WALTON, H. J. 1958. Suicidal behaviour in depressive illness. *J. ment. Sci.* 104, 884.

ALCOHOLISM

Neil Kessel

Department of Psychiatry, University of Manchester

THE condition of alcoholism commands less acceptance, less respect, from people in general, from physicians in particular, and from psychiatrists even, than its importance merits. I want to discuss three reasons for this. As a by-product I hope also to illustrate something of the picture that alcoholism presents.

Let me begin with a joke, and compliment the arrangers of this conference on their neat social engineering. They have placed a talk on alcoholism just before lunch, so that, if we are so minded, we may progress from the lecture hall to the bar, from the theoretical, as it were, to the practical, smoothly and without effort. Now, that pleasantry, if I have contrived it properly, has all the ingredients of the typical introduction to the topic of alcoholism. It was more laboured than apposite; it managed to equate alcoholism with drinking, and it thereby conveyed the shifty implication that we had better not consider alcoholism an illness in case each of us might be tarred with the brush. Lastly, without being at all funny, it yet succeeded in reducing the subject to the level of a chuckle—a snigger even. This kind of treatment of alcoholism is usual. It is eased out of serious consideration with a smile. The alcoholic is confused with the drunkard and the antics of the inebriated are good for a laugh. Such an attitude is encountered so generally that it requires exposure. Alcoholism warrants, and of course, so do alcoholics themselves, being observed with an unimpassioned regard, without emotion. At a symposium with our title we can remember Yeats' adjuration to:

> Cast a cold eye
> On life, on death.

For alcoholism presents a serious problem on many counts: it is widespread and common, it is so socially disruptive, it

130

causes suffering and disablement, both physical and psychological, and it kills.

We cannot say just how widespread alcoholism is in Britain. Prevalence surveys pose great difficulties. The problem is closely linked to the question of definition. To some people— and this is what you find in medical dictionaries—alcoholism is the result of abnormal, excessive drinking. It thus acquires a quantitative definition, volumetric and temporal. ' So-and-so was an alcoholic because he drank a bottle of whisky every day for thirty years.' This sort of definition was expressly used by Mulford and Miller [12] in a survey of a large, representative sample of Iowans: ' alcoholism is a special form of drinking behaviour '. They decided that alcoholics could be determined on the basis of personal statements about the quantity drunk, the aims of drinking and the pattern of drinking. They recorded [13] a rate of alcoholic drinking for adults of 3 per cent with a male: female ratio of 5·5 : 1.

This kind of approach takes a limited view of alcoholism. The American Medical Association Standard Nomenclature places alcoholism as a psychological illness, to wit: a personality disorder of the ' sociopathic variety '. Keller,[8] on whose essay I am heavily drawing, points out that in this classification it becomes ' a disease of no definitely known aetiology. Its origin is presumed to be in the psyche . . . obviously alcohol is not accepted as *the* causative agent ', and he continues : ' the systematists of the Standard Nomenclature have taken the position that the disease alcoholism comes from the man, rather than from the bottle '. Nevertheless, the condition must be detected by its effects, not by its causes.

In any survey designed to identify alcoholics it is therefore necessary to detect two elements: first, the evidence of repeated excessive drinking and second, the adverse social, psychiatric or general medical consequences. One study that utilized such an approach was conducted by Bailey, Haberman and Alksne.[1] They questioned one member of each of a random sample of households in a mixed district of Northwest Manhattan. ' Have you or any member of the household,' they inquired, ' ever had any health problems because of too much drinking ? . . . any job difficulty ? . . .

any money problems ? . . . family arguments ? . . . violence to members of the family ? . . . marriage break-up ? . . . trouble in the neighbourhood ? . . . trouble with the police ? . . . any other difficulties ? ' Affirmative responses were not taken automatically to indicate alcoholism. Further details were obtained and the protocol considered by the research team. They obtained an overall rate of 19 per 1000 adults, 32 for men and 9 for women. Both the overall rate and the male : female ratio were notably lower than for Iowa.

In the New York study there was little difference in the rates for white and negro males but negro women had four times the rate of white women. Age did not affect rates much. There was a great excess of widowed men and of divorced and separated people of either sex. Higher rates were found among those with the least schooling.

Commendably, but perhaps unfortunately, this team carried out a reliability study of their findings three years later.[2] A quarter of the previously identified alcoholics did not repeat their admission of alcoholism in the second survey. On the other hand, 10 per cent of previous non-alcoholics now acknowledged a drinking problem. The authors argue, not entirely convincingly, that this shows there was considerable under-reporting in both studies.

I have dwelled in some detail on that work because it is the only important field survey to have taken into account both elements of alcoholism—the drinking and the consequences. Yet its findings, as its authors have shown, are not entirely reliable and cannot be validated. They rest, inevitably, upon subjective assessments, enough to make some epidemiologists shudder.

We cannot translate estimates for New York City to London. Cultural factors are of considerable importance in determining alcoholism rates. Nevertheless, these figures, roughly 3 per cent for men and 1 per cent for women, illustrate how large a problem alcoholism is in a metropolis not so very different. I for one do not grieve over our ignorance of the number of alcoholics here. The only practical reason for requiring to know their number is to determine the need for services, and we are already sure that, whether there are 500,000 or 50,000 alcoholics in Britain, present services, both social and medical, fall far short of requirements.

It is difficult to illustrate how seriously alcoholism disrupts lives socially. This is no place for poignant case histories, but, on the other hand, statistics can convey little idea of the misery caused. To show that it provokes a considerable amount of absenteeism, debt, accidents, separation and divorce does not tell what it is like for an otherwise capable man to find himself losing his friends, drifting from job to downward job, neglecting his accepted responsibilities, lying and deceiving, or enable us to appreciate the position of a woman nightly faced with a jealous, impotent, demanding, aggressive failure, who is dragging her into debt and social isolation as she vainly struggles to maintain her own standards of behaviour and living. We had better turn to the seriousness of alcoholism in terms of psychological and physical disability and suffering.

We know the number of alcoholics, so clinically diagnosed, who are admitted to mental hospitals in Britain. The figure is in the region of 4000 annually. This does not take into account those who go into special units or into psychiatric units in general hospitals or into private care. First-admission rates are particularly interesting. In England and Wales in 1959, 26 men per million and 6 women per million were admitted for the first time with a diagnosis of an alcoholic disorder. The corresponding figures for Scotland in 1961 were 175 men and 30 women, approximately six times as many.

Alcoholism is commonly found, and frequently goes unrecognized, in the general medical wards of our hospitals. No survey has been carried out in this country but in Australia, Green[7] took a drinking history from 1000 unselected consecutive admissions to adult medical wards of a Melbourne general hospital. One in eight of all who could respond was an alcoholic, defined as ' at some time having been physically, mentally or socially incapacitated by prolonged drinking '. This is a rather over-embracing definition but the text of the article makes it clear that simple episodes of drunkenness did not fall within its scope. Moreover, a third of these alcoholics were admitted with illnesses directly related to excessive drinking, and a further fifth had illnesses which were possibly associated.   At least 4 per cent, therefore, of all medical admissions were the consequence of excessive drinking.   A study in a teaching hospital in Connecticut [14] yielded similar results.

Besides their hospital tenancy, alcoholics make other claims on public beds.   At least a hundred chronic alcoholics are detained in London's prisons on any one night and many of them regularly reappear on charges.   It is not uncommon for one man to have been in prison fifty times.[10]   Moreover London, like most other large cities, has its skid rows in which many alcoholics nightly fetch up.[4]

A considerable number of alcoholics kill themselves.   The suicide rate for male alcoholics in London was some eighty times that for all males of comparable age.[9]   Yet suicide is only one way in which alcoholics die.   Lipscomb and Tashiro [11] showed that the average annual death rate among alcoholics in California was two-and-a-half times the normal rate.   The most frequent cause of death was violence.   But cirrhosis of the liver also takes its toll, for there are very high rates of its incidence in occupations with a high risk of alcoholism : brewers, publicans, etc.   Bandel,[3] in a series of studies (summarized by Freudenberg), demonstrated parallels, both regional and secular, between alcohol consumption per capita and male mortality in general.

A tame, lame view used to be expressed that those involved in drunken driving accidents were not alcoholics, for these were too incapable ever to get to the wheel.   Glatt [5] has shown, on the contrary, that alcoholics have frequently been ' in trouble because of driving when drunk ' : 23 per cent of his male alcoholics and 12 per cent of his female alcoholics gave such a history.   In a later study,[6] he discovered that very much higher proportions than this admitted having driven while intoxicated.

On all counts—general medical, psychiatric, the personal social level and the societal—alcoholism is a serious problem.

The second reason why alcoholism is often excluded from the canons of medicine is that its limits are ill-defined and the framework of the illness picture and its natural progress are not appreciated.   Consider, for example, the World Health Organization definition.[16]   In its efforts to please everyone and to be all-embracing, it has produced a formula that could include nearly everyone.

> Alcoholics are those excessive drinkers whose dependence on alcohol has attained such a degree that they show a noticeable mental disturbance or an interference with their mental and bodily health, their interpersonal relations and their smooth social and economic functioning; or who show the prodromal signs of such developments.

Certainly the elements of the condition are there but, even without that final rider, they can be so variously interpreted that they are useless in delineating the condition. " Don't you have any test for it? " I was plaintively asked recently. Alas, we do not. But it is none the less possible to indicate the outlines of the condition—to give it a structure within which to fit the progressive clinical features that we meet.   There are three relatively distinct stages and nothing pleases physicians more than that.

First, there is the stage of excessive drinking with social problems, marked by the evident occurrence of those very things.   The alcoholic drinks more, sneaks drinks, becomes pre-occupied with drinking.   Physiologically, his tolerance for alcohol rises.   Psychologically, he may find himself drinking to get relief from tension—he uses alcohol almost as a medicine. He feels he needs drink to perform adequately at work or socially.   At the same time his work begins to suffer and his friends begin to notice that he is always to be seen with a drink.   He in his turn becomes aware of their critical scrutiny and a vicious circle develops.   Absenteeism, often with the loss of job, and social withdrawal follow.   Debts mount and family tensions inevitably develop.

The second stage is generally called the phase of alcohol addiction.   The alcoholic can not now give up drinking simply by his own efforts, though it is rare for him to accept this. Though the social decline continues this stage is dominated by the occurrence of psychological abnormalities.   Early on, at least, these are fortunately reversible.   Alcoholic amnesias occur.   He calls them ' blackouts ' because in the morning he wakes with no clear recollection of the events of the previous

K

night. This is not because he became unconscious through drinking; in fact he took an active part in events. But no memory of them remains. Round about this time, too, there is a loss of self-esteem and the development of remorse. Soon, however, something new occurs. So far he has blamed himself; now he begins to regard himself as the victim of circumstance. He used to reprove himself for coming home late, drunk; now he feels that he stays out because his wife nags, and he drinks to fortify himself against her nagging. He convinces himself that it is because the boss is unfair that he loses his job. The onset of these paranoid misinterpretations (which unfortunately may contain a grain or two of truth) is an important milestone. Morbid jealousy—the conviction that one's spouse is unfaithful—falls into the same category.

Another milestone is the development of withdrawal symptoms if he stops drinking—tremulousness, transient hallucinatory states, sometimes epileptic fits, and also the striking florid delirious state of delirium tremens, are all psychological danger signs that the brain of the alcoholic is being damaged. Despair prevails and there are commonly episodes of deliberate self-poisoning or self-injury which carry, as we have noted, a high risk of suicide.

The final stage is that of chronic alcoholism. There is continuous drinking, with reduction in tolerance, so that thinking is confused for prolonged periods. This stage is characterized by the development of severe physical illnesses, both of the brain and of the rest of the body. These are partly toxic in origin and partly the result of vitamin deficiencies occasioned by malnutrition. The prognosis is now grave. It hangs between kill or cure and the chances of cure are not good.

There is little cogency in either of these reasons why alcoholism so evades serious consideration. In the first place it is not a joke. Secondly, it can be described and defined, and its stages and progress codified in the same way as other illnesses. But there is a third reason why doctors in particular are sometimes reluctant or unready to consider it as an illness. And this reason has wider ramifications than relate to alcoholism only, and its implications are increasing.

The alcoholic bears some responsibility for his condition.

To consider illness as something for which the patient has responsibility makes doctors uneasy. This is not what they were taught as students, nor did they find it anywhere written into the Hippocratic Oath. On the contrary, the view was consistently put forward that illness was something that happened to people without their willing it in any way, and that they were powerless to prevent. They went to doctors and put themselves in their hands to be made well again. Passively they became ill; passively they were healed.

During the course of their education medical students found, to be sure, dark sectors of illness where this traditional approach was less easily tenable. Venereal disease was one, but the VD clinic was generally pushed to a small back basement of the hospital and held at awkward hours. Students had little contact with the department, and may have been forgiven for thinking that its dinginess somehow implied an attitude of just retribution or even of punishment on the part of their teachers.

As knowledge concerning disease prevention spreads, however, ordinary people can increasingly reduce the risk of contracting certain illnesses; if they do not, then they bear some responsibility for getting them. Nowadays, a man must shoulder a major share of responsibility for catching smallpox. This is obvious. He could effectively have protected himself. A man who begins smoking now and later develops carcinoma of the bronchus also bears some of the responsibility. So does a man who persists in remaining overweight and who develops, say, diabetes.

Doctors will have to learn to accept this. It involves accepting a patient's responsibility without reacting to the patient by censure or in a punishing way or to the illness by excluding it from his ambit. At the moment the doctor still acts retributively towards his alcoholic patients. You should read some of the harsh strictures contained in letters with which general practitioners often refer alcoholics to clinics, if they do refer them. Consider some of the answers they gave to Parr [15] when he inquired about alcoholics in their practices. One doctor wrote: " Generally speaking, I suggest alcoholics avoid doctors and doctors in the main try to avoid alcoholics."

But how far is it correct to regard the alcoholic as responsible for his illness? We have to consider the causes of alcoholism, and we do not know them at all well. I have already referred to the dispute as to whether it springs from the man or the bottle. The complicated relationship between these two is summed up pithily in an old saying: First the man takes a drink, finally the drink takes a man.

Certain features occur more often in alcoholics than in others, and *probably* they antedate the illness. Particular personality structures—not necessarily abnormal—are commonly found: immature personalities, self-indulgent personalities, people beset with sexual problems, self-punitive individuals. Of course the majority of people with these personalities do not become alcoholics, nor do all alcoholics have such personalities. They are neither sufficient nor necessary causes, but they are aetiologically important nevertheless.

Social influences are also involved, acting through example, opportunity and incitement. Example comes from one's parents and one's peers, incitement from one's peers, from advertising and perhaps also—here I am thinking of the high alcoholism admission rates in Scotland—from the lack of varied, alternative leisure pursuits. Or is it the number of people engaged in the drink industry in Scotland which accounts for the difference? Opportunity comes often from one's job, from one's wealth and also from society's laws regarding the sale of drinks and from cultural pressures to drink on various social occasions. But again, most of the people who are set a bad example, who are incited to drink and presented with abundant opportunity, do not become alcoholics, so that no particular social influence is a sufficient cause, either.

However, all attempts to find causes of alcoholism outside personality and social influences have failed. There is no satisfactory evidence that physical or genetically inherited factors play a part in determining that a man should become an alcoholic.

Therefore, since we generally regard a man as responsible for the fruits of his personality and also for responding in a deviant way to social influences, we must regard him as bearing some measure of responsibility for his alcoholism. Indeed, everyone, if he is honest, knows this; certainly both alcoholics and their families do. It is, in fact, largely the medical profession which finds it difficult to accept both this and that alcoholism is an illness.

There is an increasing number of other illnesses for which a person carries responsibility whether he gets them or not. Alcoholism is but one example but I hope I have shown that this third reason for not respecting alcoholism as an illness is also untenable. We require therefore to reappraise our attitudes to the whole question of responsibility for illness. This Society could be in the vanguard in such a change.

REFERENCES

1. BAILEY, M. B., HABERMAN, P. W. and ALKSNE, H. 1965. The epidemiology of alcoholism in an urban residential area. *Quart. J. Stud. Alc.* 26, 19.
2. BAILEY, M. B., HABERMAN, P. W. and SHEINBERG, J. 1966. Identification of alcoholics in population surveys: a report of reliability. *Quart. J. Stud. Alc.* 27, 2.
3. BANDEL. 1931. Summarized by FREUDENBERG, K. *Klin. Wschr.* 10, 106.
4. EDWARDS, G., WILLIAMS, G., HAWKER, A. and HENSMAN, C. 1966. London's Skid Row. *Lancet* i, 249.
5. GLATT, M. M. 1961. Drinking habits of English (middleclass) alcoholics. *Acta psychiat. scand.* 37, 88.
6. GLATT, M. M. 1964. Alcoholism in 'impaired' and drunken driving. *Lancet* ii, 161.
7. GREEN, J. R. 1965. *Med. J. Aust.* 1, 465.
8. KELLER, M. 1962. The definition of alcoholism and the estimation of its prevalence. In *Society, Culture and Drinking Patterns.* (Ed. D. J. PITTMAN and C. R. SNYDER.) New York. Wiley.
9. KESSEL, N. and GROSSMAN, G. 1961. Suicide in alcoholics. *Brit. med. J.* ii, 1671.
10. *Lancet.* 1965. i, 253.
11. LIPSCOMB, W. R. and TASHIRO, W. 1963. Mortality experience of alcoholics. *Quart. J. Stud. Alc.* 24, 203.
12. MULFORD, H. A. and MILLER, D. E. 1960a. Drinking in Iowa. III. *Quart. J. Stud. Alc.* 21, 267.

13. MULFORD, H. A. and MILLER, D. E.   1960b.   Drinking in Iowa.   V.
        Quart. J. Stud. Alc. 21, 483.
14. NOLAN, J. P.   1965.   Amer. J. med. Sci. 249, 135.
15. PARR, D.   1957.   Alcoholism in general practice.   Brit. J. Addict. 54, 25.
16. WORLD HEALTH ORGANIZATION.   1952.   Expert Committee on Mental
        Health.   Alcohol subcommittee: Second Report.   Tech. Rep. Ser.
        Wld. Hlth. Org. 48.

CAUSES AND EFFECTS OF AGEING

Chairman: PROFESSOR E. MAURICE BACKETT

INTRODUCTION

E. Maurice Backett

Department of Public Health and Social Medicine, University of Aberdeen

In this, the last Session, we deal with the causes and effects of ageing. In our discussions we have already met many of the causes, a few of the effects and some of the problems, and a little of the strictly demographic background has also been mentioned. It seems that in many (but not all) of our communities more and more people are surviving to substantial old age. As the proportions change, so do our standards and what we expect in terms of comfort, health and happiness alter from year to year. It is with the immensely wide and altogether fascinating implications of these changes that we are now concerned.

Survival to old age does not, of course, imply a greatly increased expectation of life for the aged. By far the most important part of the social patterns with which we are concerned is the greater number of old and very old people, rather than with a longer life for us all. So, while at its crudest this is a problem of numbers and proportions, it is much more than that; for the sheer size of the problem is forcing us, whether we like it or not, into a re-appraisal of the quality of life lived by the elderly.

It is only in the surprisingly recent past that medicine has made any real impact upon the health of populations. Before that, little could be offered beyond sympathy—always an important ingredient of medical care. Perhaps because medicine was so powerless, the main challenge has always seemed to be that of premature death. When at last doctors began to have some power over disease they continued (perhaps illogically) to put first things first; their preoccupation remained with death control. They can be forgiven therefore for being less concerned about the quality of the life they saved, for this is rarely seen as a strictly *medical* problem and anyway is almost

143

always surrounded by controversy. In contrast the issue of survival or not survival has an attractive clarity.

The prevention of death—especially premature death—is obviously important, but the time has now come for more careful scrutiny of what we mean by ' premature ', and of the social, economic and pathological features of the lives of our very old people.

In spite of much recent research, our information is still poor in quality and relatively small in amount. At the vital statistical level the multiple pathology of the aged makes a large proportion of death certificates valueless. As a result—and very much to our cost—we omit the older age groups from our epidemiology, and though much is written about the morbidity and social needs of the aged we seem to make some terrifying mistakes in social policy. Research workers are struck by the extraordinary ease with which it is possible to help at the personal level, and to do serious damage through our ignorance at the community level—for example, our confusion over the classification of institutional care. Accurate information about needs and demands is too often lacking.

In the study of ageing there are a number of untouched epidemiological goldmines; one example is the study of the experience of cohorts subjected to differing social environments. This kind of study, like so many others, requires the meeting of medicine and the social sciences. In fact, the study of old age is an ideal meeting place of disciplines, a place where geneticists and epidemiologists, sociologists and biochemists, demographers and clinicians will together contribute to our understanding of the biology of ageing. It is therefore as a multidisciplined exercise that we must approach these problems and also the papers that now follow.

SOME ECONOMIC AND SOCIAL
CONSEQUENCES

PETER R. COX
Government Actuary's Department

I T will perhaps be appropriate if this introductory talk harrows lightly over a wide area instead of ploughing deep in any one field, and that is why the title chosen is broad but not comprehensive. This title takes the order of its words for reasons of euphony, and so indeed does the title of the whole Session. It so happens, however, that in each case the order in which the topics are considered is to be changed, so as to get into the subject in the best way. Thus, we are going to speak of the consequences of ageing before the causes; and social consequences will be interspersed among the economic ones: in any event there is no very clear-cut distinction between the two.

AGEING OF POPULATIONS

When we investigate the causes of ageing, we are dealing with something which operates on and through individuals, but when we speak of effects we may be thinking of the outcome either for a particular person or for a group of people. The distinction between the individual and the group in this connection is important and failure to recognize it has sometimes led to confusion of thought. I propose to begin, therefore, by discussing what we mean when we say that a population is ageing. A person aged sixty-five has obviously lived longer than one aged twenty-five (I do not say that he is ' older ' because one is only as old as one feels). Comparison of two populations for their agedness is, however, a problem in statistics. In accordance with well-established procedures, we might sum up by saying that population A has a certain mean age and a certain standard deviation. We might also make

145

some reference to the higher-order moments of the distribution, or to its quartiles or deciles. Then we could set out these summary measures alongside the corresponding measures for population B and we should have a clear and reasonably comprehensive comparison; but this is a method of presentation that demographers, generally speaking, do not use. Instead, they tend to speak of the proportion of persons over pension age, the proportion who are children at school, and so on, although these are less effective functions of measurement. They do so, I think, because demography is often the servant of sociology and of economics, and in both capacities is heavily dominated by the roles that people perform at different, fairly definite, stages of their lives.

On any basis of measurement, however, it is clear that a number of national populations have aged in recent years or are in process of ageing. This is not merely because individuals grow old; for, in spite of that, it is possible in certain circumstances for a population to become progressively younger over a period of time. The ageing of a population depends entirely on the comparative rates at which three things happen:

    i. infants are added to the population by birth;
    ii. people, usually young, are added or subtracted by migration; and
    iii. people of all ages are subtracted by death.

As there is still a good deal of misconception about this, I propose to discuss it a little further before proceeding to other subjects.

Figure 1 shows the age-distributions of the population of England and Wales in 1871 and 1921 and according to the latest projections for 1971 and 2001. It shows that at the younger ages the proportions fall steadily to the 1971 level and that at the older ages they rise uniformly. From 1971 to 2001, however, the process is reversed. This point is worth noting, because it is not yet generally realized that the ageing of our population has now virtually come to an end. The numbers of the aged will continue to grow significantly, but their proportion should fall a little before long.

Although migration can sometimes play a significant part in affecting the age-distribution of populations, in general its contribution is small because only rarely is it physically possible for more than a tiny fraction of the population to move across international borders or seas within a period of time.

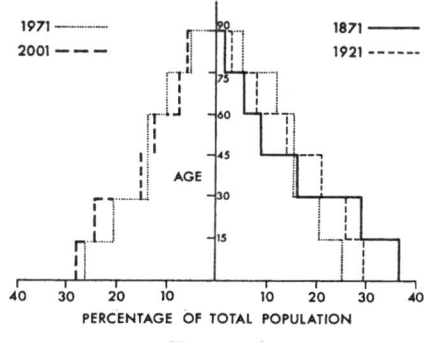

FIGURE 1

Age-distribution of the total population of England and Wales in 1871 and 1921, and as projected for 1971 and 2001

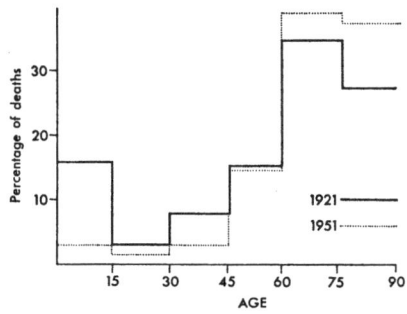

FIGURE 2

Age-distribution of men's deaths according to *English Life Tables No. 9* (1921) and *No. 11* (1951)

The effect of improvements in mortality upon the age-distribution of the population is not what might be expected: an increase in the expectation of life at birth does not necessarily lead to an older age-distribution for the population as a whole.

Figure 2 shows the percentages of men's deaths occurring

in successive fifteen-year age-groups according to the national life tables of 1921 and 1951 for England and Wales. It will be seen from this that young lives have been saved and older ones lost, which does not suggest ageing of the population. The comparison needs some correction because the age-distributions of the living persons in the 1921 and 1951 life tables are not the same.

To eliminate this objection, let us apply the death rates of

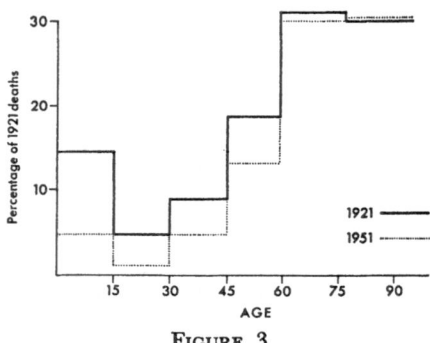

FIGURE 3

Age-distribution of expected deaths in England and Wales based on 1921 mortality and a standard population, and corresponding distribution based on 1951 mortality. Men.

1921 and those of 1951 to a standard population—say that of men in England and Wales in 1957—as in Figure 3. This improved comparison shows a saving in lives at all ages up to seventy-five, but most of all in youth. The average age of the lives saved is, in fact, only twenty-seven, which is *below* the mean population age. The effect of mortality improvements over this period was therefore to make the population younger; but the effect was not a very pronounced one.

As neither migration nor mortality has had an important effect on ageing, it must then be changes in fertility that have brought it about. That this is so has been conclusively proved for many western countries over recent periods. I will not go into all the details but will confine my remarks to a simple illustration.

In Figure 4 is shown a life-table population, in which a constant supply of births is just sufficient in number to maintain the population at a constant level. This is indicated by the continuous line. The broken line shows what happens if this is adjusted so that the population is supplied by enough births to double it every thirty years, the mortality rate being unchanged. This has a big effect in reducing the age of the population. The difference is commensurate with disparities

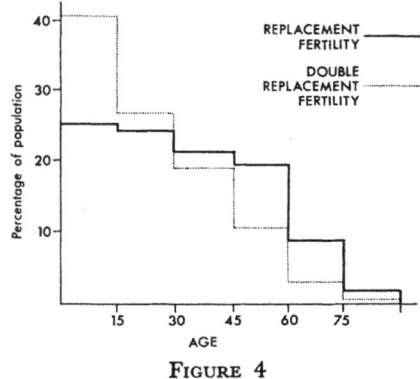

FIGURE 4

Age-distribution of hypothetical populations with the same mortality and (i) replacement fertility; (ii) double replacement fertility

in age-distribution, and in fertility, between western populations and populations of emergent countries in Africa, Asia and South America.

When the age-distribution of a population changes, and particularly when it develops steadily over a period, as has happened in recent decades in Britain, it is of some interest to measure the ratio of the size of the ' dependent ' groups in the population to that of the active and producing groups. Exercises of this kind have been undertaken in recent years, and some writers have even attached differing production and consumption ' weights ' to different age-groups in an attempt to strike an economic balance.

Figure 5 gives, without any weighting, the ratio of the old to the active, showing the steady and substantial rise followed by a slight fall, and the ratio of the children to the active, showing the opposite tendency. The curve for both classes of dependant together is naturally flatter but is now showing an upward track after a dip.

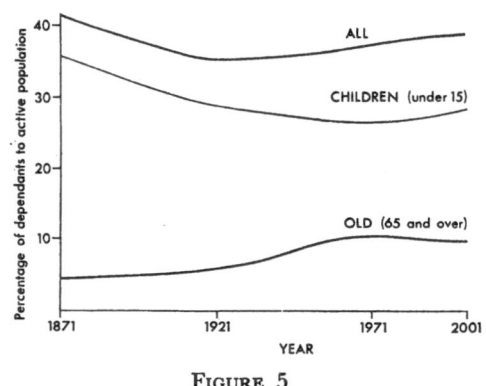

FIGURE 5

Percentage of dependants to total population.
England and Wales, 1871–2001

CONSEQUENCES OF AGEING

Now that the proportion of the aged has stopped rising, and some regular annual growth in national wealth can reasonably be expected, the emphasis placed upon the burden of dependency has greatly diminished in recent writings. During the 1950s it gave way to a general consciousness that the cost of old age should be quite bearable, if the level of economic provision was reasonable. At that stage, however, much attention was paid to the considerable fiscal and other financial problems involved in making the transfer payments from the active to the dependent groups.

Because of the existence of those problems some have suggested that a selective rather than a wholesale transfer of resources would be not only more manageable but also more just. In the 1960s, therefore, the object of research has often been the ascertainment of differences in need from one person to another and from one group to another. I propose to describe some of the researches of this kind that have been made in recent years.

The consequences of ageing are well known. There is, for instance, an increased liability to illness and an enhanced risk of unemployment, coupled with the virtual certainty of retirement at some stage or other. There is the likelihood of a reduced income during retirement and perhaps even before it. The chance of separation from the rest of the family, such as that brought about by widowhood, leading perhaps to loneliness and isolation, is increased.

Many inquiries have been directed to discovering the best way in which the community as a whole can cope with these problems. The liability to an increase in sickness involves a higher cost of social services and a growing demand upon the doctor's time, as well as a greater need for family care. When separation occurs, another requirement for care is created, and if this care is not available from the family, perhaps special housing or some form of institution may be called for. Retirement may be forced too soon by social and economic pressures, contrary to the welfare of the individual, whose need is to work as long as he possibly can in the right type of employment. When retirement comes, individuals are often unprepared and fail to adapt properly to it. When the amount of their income falls there are various ways in which it can be supplemented; the possibilities are varied and offer scope for frequent investigation.

In response to such problems we have the Preparation for Retirement movement, about which I would like to say a few words later. We have had investigations into the opportunities that exist for the employment of older people, and into the extent to which working life is in practice prolonged. The respective roles of the family and the community in caring for the aged have been the subject of particular investigations. The available methods of providing resources for older people have been re-examined, in order to find out whether the system of social services devised twenty years ago is appropriate to-day. Cases of failure to apply for State aid have been the

L

subject of official concern and study.   It is to such activities
that I now draw attention.

RETIREMENT

It is of some interest to note the extent to which people do
in fact soldier on, and Figure 6 shows the position for men of
all employments in England and Wales in the two years 1931
and 1955.[2]   Of 100 men who reached age sixty, it indicates
for successive ages the cumulative numbers of those who died
at work, who retired and who are still active.   It has to be
borne in mind that these are only two individual years and
may not fully represent the inter-war and post-war periods
respectively.   In particular, 1931 was situated at the beginning
of the depression, and on the other hand we have had considerable economic developments since 1955.   It must be
remembered also that the extent to which employment is
prolonged in old age will vary a great deal, not only from one
occupation to another but also from one status to another, and
indeed from one individual to another.   For what the figures
are worth, however, they show that it was easier in 1955 than
in 1931 to stay on at work until age sixty-five, but between
sixty-five and seventy, and indeed immediately at age sixty-
five, retirement was generally faster in 1955 than in 1931.
Some data for the United States of America in 1955 [7] indicate
an even more rapid rate of departure from the labour force
between sixty and seventy than in Britain, and this may of
course be a pointer for our future.

In 1954 the Ministry of Pensions and National Insurance
conducted an inquiry [4] into the reasons people had for retiring or continuing at work.   Men reaching age sixty-five and
women reaching age sixty become entitled to national insurance pensions if they retire at those ages.   If they do not retire, then the pensions (augmented) are payable on retirement
if that occurs within five years, or in any event at age seventy
for men or sixty-five for women even if retirement has not then
taken place.   The Ministry asked persons retiring at the
minimum age, or at some time within the following five years,
why they had retired.   They also asked those not retiring

at the minimum age, and those still active five years later, why they had continued at work. The distribution of people according to whether they went at the minimum age, stayed on for five years or retired in the interim is broadly indicated

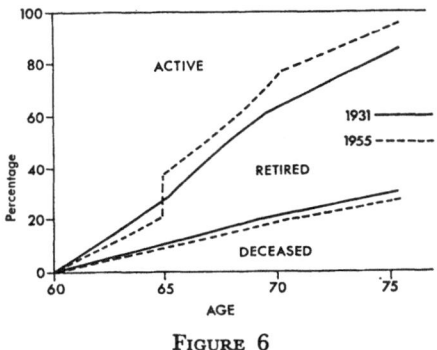

FIGURE 6

Percentage of men active, retired or deceased after age 60. Britain, 1931 and 1955.

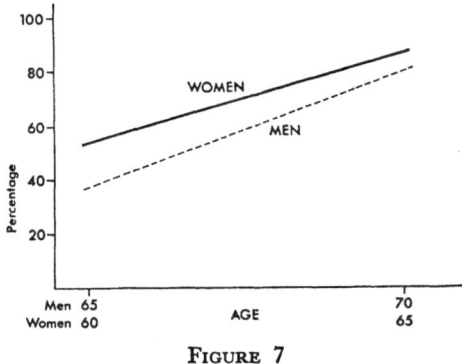

FIGURE 7

Percentage retired among men and women.
Britain, 1954.

in Figure 7. The figures do not indicate quite the same picture as Figure 6 did for men in 1955. This is because of a difference of approach. A substantial body of people are already chronically sick when they reach minimum pensionable age. In the inquiry of the Ministry these were included as retirements at the minimum age, because they thus moved from sickness benefit to pension benefit, whereas in Figure 6 they were classified as persons already out of the labour force before pension age. Very broadly, ill-health was given as the cause of retirement in about one-half of the retirements. Part of this represents, as I have indicated, those who had already been permanently sick for some time, but in addition to these there were many who did not feel well enough to go on working. Those who were not ill, but were discharged by the employer, accounted for about one-half of the remaining retirements of men and much less than one-half for women. The remainder gave a miscellany of reasons for retiring. Those who stayed on, on the other hand, often gave financial need as a cause of continuance at work, and this is highlighted by the fact that those without the benefit of any occupational pension were more inclined to stay on than those who had both a national and an occupational pension. About one-half mentioned financial need as a reason for continuing at work. Most of the remainder said that they felt fit or preferred to go on working. These people make it clear that, except in cases of ill-health, there is plenty of desire to stay at work, and reason for it too. More recent surveys have confirmed this desire, coupled with a realization that a change of job to lighter or part-time work might be advisable.

In modern conditions, however, suitable work may not be available, as the studies of le Gros Clark [1] have shown. The conditions of the mechanized world in which we live weigh heavily against the older man. They do so by applying, in effect, more vigorous tests of ability than did working conditions in the past and also by limiting the range of lighter work suitable for the elderly. Light work is often given to trainees in preference. Clark felt that this unfortunate situation is only a temporary phase and should right itself in the future. He also felt, however, that a new social theory of retirement is needed, including perhaps not only the better use of leisure but also more productive activities. Although his writings are persuasive, the evidence for the development of a new social theory of this kind seems scanty at present.

Recent comprehensive inquiries by Townsend and Wedderburn,[5] and by the Ministry of Pensions and National Insurance,[3] in which old people were interviewed at their homes, have provided us with a clear picture of certain aspects of the life of the aged in Britain to-day. Their living arrangements, their family connections and support, their health and mobility, the help they get from the various social services, the help they would like but do not get, and the help they could have but reject, have all been measured.

HEALTH AND MOBILITY

I should perhaps make some reference to the effect of increasing age on health, although this brings me rather near to the field covered by Mr R. D. Clarke. I shall not show any figures on this subject, however, because there is the difficulty that the criterion of fitness must necessarily change as the age advances. In one official inquiry, illness meant inability to work for those still in employment, inability to leave the house for those retired but still mobile, and inability to leave one's bed for those normally confined to the house. It is, indeed, difficult to think of any useful single criterion which is equally applicable to all such circumstances. In their new inquiry the Ministry of Pensions and National Insurance asked pensioners (other than those in hospital) for subjective assessment of their state of health, and the results are interesting. The proportion who stated that their health was good dropped a little, and the proportion who said that their health was poor or very poor rose a little, with advancing years. The extent of the change was, however, only moderate. Of those aged eighty-five and over one-quarter claimed to be in good health and 40 per cent regarded themselves as being in fairly good health.

In the experience of friendly societies which pay sickness benefit at all ages, the proportion claiming benefit rises steadily to 100 per cent at the most advanced ages, and the number of weeks for which sickness lasts in the course of a year increases till it reaches 52. This is a very different picture, but the disparity from the subjective assessments probably reflects two fundamental principles :

    i. Pensioners' expectations of ' normal ' health probably

change as their age advances, and their assessment of their fitness tends to deviate more and more from the view of an independent medical expert; and

ii. Increasing financial needs, and a greater ease in establishing title to sickness benefit, are experienced in very old age.

As the years pass, mobility tends to become more and more restricted. Townsend and Wedderburn found that about 2 per

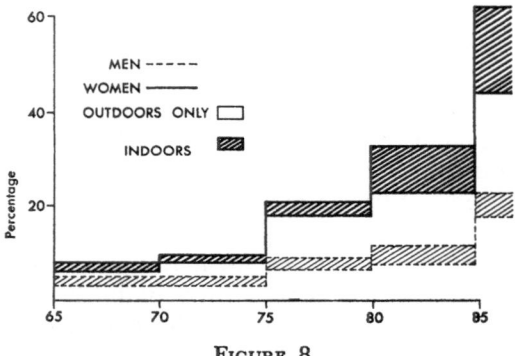

FIGURE 8

Percentage of old persons unable to get about
unaided.   Britain, 1965

cent of those aged over sixty-five and not living in institutions were bedfast, that 11 per cent were confined to the house, and a further 8 per cent could go outside only with difficulty.   For those in institutions the figures were even more unfavourable. Some figures collected by the Ministry are illustrated in Figure 8, which demonstrates for single pensioners not in hospital how rapidly the proportion of incapacitated persons increases for both sexes after age seventy-five.

With restrictions on mobility is associated the loss of one faculty or another and an increasing inability to attend to matters which help to keep one in good health.   Townsend and Wedderburn found that the percentages of persons aged sixty-five and over suffering particular disabilities were as indicated in Figure 9.   This shows how large hearing and seeing difficulties loomed; while the need for chiropody assumes a surprising degree of importance—apparently it makes all the difference between mobility and immobility for very many people, and in old age mobility is important for general health, and especially mental health.

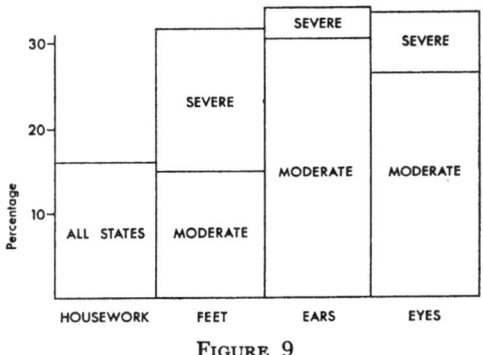

FIGURE 9

Percentage of old persons experiencing difficulties of various kinds. Britain, 1964

HOUSING AND INCOME

From questions of disability it is not a far cry to the consideration of household living arrangements. The Ministry found that the proportion of married couples who were householders was very high—98 per cent, and that as many as 65 per cent of single men and 76 per cent of single women had their own place. Most of the remainder lived with relatives, but 12 per cent of single men and 3 per cent of single women were boarders or lived in institutions. Figure 10 shows how the proportion living alone falls away as age advances, although it is interesting to note that in the early years of old age there is actually a small rise in the ratio for men. Townsend and Wedderburn gave a table showing what relatives people live with: it shows that the main classes are their married children, unmarried children, and siblings, in that order. The rise in this ratio may therefore be due to the death of siblings, the marriage of children, or an increase in the size of the family of married children.

I will not stress the question of the ' aloneness ' of the old, as it seems likely that most old people prefer to be alone. In an American discussion, the word ' privacy ' was used, and this

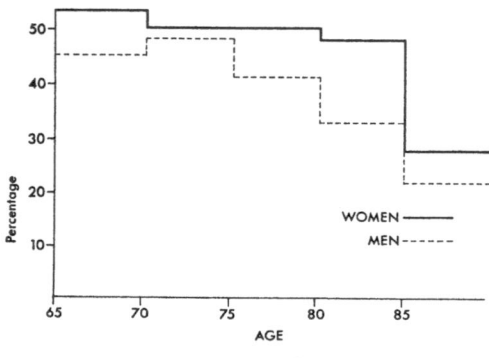

FIGURE 10

Percentage of old people living alone.
Britain, 1965

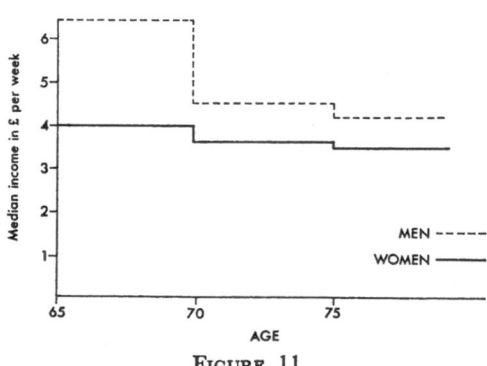

FIGURE 11

Average incomes of single men and women at
advanced ages.   Britain, 1964

may convey rather a better notion of the position than something which implies loneliness.

Figure 11 illustrates what Townsend and Wedderburn said about the size of incomes as old age increases. The figures include people still at work, but if the data are restricted to retired people alone a similar picture is presented, although the income level throughout is lower.

The Ministry found that resources exceeded needs, assessed in terms of national assistance entitlement, in 71 per cent of married couples, 65 per cent of single men and 45 per cent of single women.   In the remainder, needs exceeded resources, by the national assistance criterion, although in many cases disregarded income would in fact bridge the gap; but ignoring this, in more than one-half of such cases the gap was being

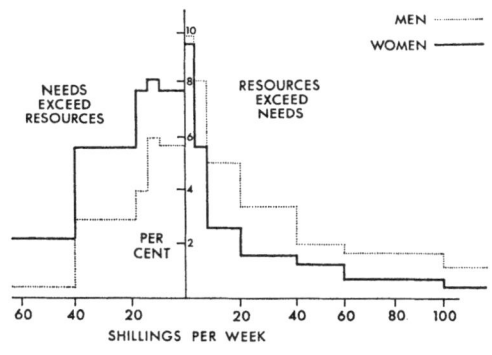

FIGURE 12

Differences between amounts of needs and resources; single men and women at advanced ages. Britain, 1965

bridged by national assistance.   The extent of the gap before taking credit for any national assistance is illustrated in the left-hand half of Figure 12.   Often the gap is quite small but there is also an appreciable tail to the distribution, rising to £2 a week or even more.

The unofficial and the official inquiries to which I have referred cover the same ground to some extent, and where they do the results are encouragingly consistent.   There is, however, an interesting difference of emphasis.   Townsend and Wedderburn say that they were concerned to find out the extent to which the welfare services were falling short because they did not reach people who felt a need for them.   Although they did show the cases where the need was being successfully met, they rather concentrated on the failures. The Figures given above show the position for the total need of one kind or another, whether or not it has been met by the social services. The Ministry of Pensions and National Insurance, however, were concerned to find out how many people were unaware of the possibilities of State aid and therefore failed to apply for it when they could reasonably do so, and how many people knew about the State aid but refused to ask for it, and why they did so. This approach is to some extent only the other side of the same coin ; but the fact remains that the coin has been looked at from different angles in the two reports. For my purpose this has been useful, because the picture as a whole can be seen in a clearer light. It should be borne in mind, however, that if one was asked on one's doorstep what one felt one needed there would be some temptation to let one's fancy wander and think of more things than one would be willing to apply for in practice, particularly in the light of experience of what the services can actually do. The official inquiry with its pragmatic approach is evidently of great value. At the same time, the investigations of Townsend and Wedderburn are of particular significance in that they underline the continuing importance of family help in Britain to-day in both cash and kind and in personal service too. State aid, they find, has by no means undermined the principles of family welfare and most of the problems to which they draw attention are those of people who no longer have any family to turn to.

Recently Tunstall [6] has referred to the comparative neglect of ageing in social theory, which has directed its attention primarily to the young. I agree with his plea for research, but think I can see why the subject has not been developed. Is it not that ageing represents to some extent a negation of society ? So long as old people are able to help themselves, perhaps we should concentrate on advising them how best they can do so, especially in retirement.

OTHER PROBLEMS

Some of you may have seen on television parts of a useful and interesting series of programmes entitled ' Forward to

Retirement '. The series was prepared by the BBC in collaboration with a body known as the Pre-Retirement Association. This Association was formed in 1964 as the natural outcome of studies initiated by the National Old People's Welfare Council in 1955. The primary object of the Association is to alert working men and women to the fact that retirement and the new pattern of living that it involves are as worth preparing for as much as any other phase of life. It is a servicing body, providing a clearing house for the rapidly accumulating information on, and experience of, the way in which such preparation can be made. Though it has not yet achieved the necessary financial support, there is no doubt of the need for a central body to cope with the increasing awareness that retirement presents a social problem which the community must face. I understand that some firms undertake their own programmes and arrangements with a preparatory aim, and if this is properly done, it is all to the good.

The elements in which ordinary people seem to need education to prepare for retirement are personal relationships, finance, health, housing and hobbies. The emphasis placed upon personal relationships may seem a little surprising; but couples are not always ready for the difficulties that can arise when the husband is at home all day getting in the way instead of only in the morning and evening. The importance accorded to hobbies is less surprising; I never fail to be impressed with the variety of useful and fascinating leisure pursuits available and what some people can do with them, but I am equally surprised how few people, relatively speaking, take up such hobbies. There can be no doubt that education for leisure is sorely needed and that we have a lot to learn about teaching methods in this subject. It appears, however, that unlike the matters I have already mentioned, this question has been the subject of very little research. This seems to be an area in which investigation would be valuable, and I certainly hope that before long it will engage the closer attention of sociologists and educationalists. Perhaps this will be the next major change in direction in sociological research into ageing.

REFERENCES

1. LE GROS CLARK, F.   1956.   *The Employment Problems of Elderly Men.* London.   The Nuffield Foundation.   (And other works by this author, published by the Foundation.)
2. MINISTRY OF LABOUR AND NATIONAL SERVICE.   1959.   *The Length of Working Life of Males in Great Britain.*   London.   H.M.S.O.
3. MINISTRY OF PENSIONS AND NATIONAL INSURANCE.   1966.   *Financial and other Circumstances of Retirement Pensioners.*   London.   H.M.S.O.
4. MINISTRY OF PENSIONS AND NATIONAL INSURANCE.   1954.   *Reasons Given for Retiring or Continuing at Work.*   London.   H.M.S.O.
5. TOWNSEND, PETER and WEDDERBURN, DOROTHY.   1965.   *The Aged in the Welfare State.*   London.   Bell.
6. TUNSTALL, JEREMY.   1966.   *Old and Alone.*   London.   Routledge.
7. WOLFBEIN, SEYMOUR L.   1957.   *The Length of Working Life.*   Merano, Fourth International Gerontological Congress.

PSYCHIATRIC ASPECTS OF OLD AGE

MARTIN ROTH

*MRC Research Group on the Relation between Functional
and Organic Psychiatric Illness
Department of Psychological Medicine, Newcastle-upon-Tyne*

OWING to the swift rise in the proportion of old people in the population of most economically advanced countries and the successful prevention and treatment of many forms of illness in early life, the disabilities that bear some relationship to ageing have been thrown into increasing prominence. It is logical to give consideration to mental disorder in this context in that, if first admission rates to mental hospitals are taken as an index, the expectation of psychiatric illness shows a steep rise after the age of fifty years in both sexes. Psychiatric institutions in economically developed countries have therefore had to deal with ever increasing numbers of aged people in recent years. But the change is too great to be explained in terms of ageing of populations alone. There has probably been no increase in the rate of first admission of patients with major psychoses to mental hospital during the past century.[6] Moreover, according to the most widely held theories about the genetic basis of these disorders, it seems possible that their incidence during the first fifty years of life has varied over many generations only within relatively narrow limits. Most societies therefore have a long, if not always untarnished, experience of dealing with schizophrenic and depressive illness. On the other hand, a hundred years ago the expectation of life at birth was forty years and the proportion of individuals aged sixty-five and over in the general population of this country has almost trebled since the beginning of the century. The scale of the problem of mental disorder among the aged and of the physical diseases closely associated with it is unprecedented.

Many questions in relation to mental disorder in the aged are prompted by a symposium such as this. To what extent are we dealing, in the psychiatric illnesses of old age, with conditions arising from senescent degeneration of the brain, and how far with disorders independent of such changes? How far are we dealing with hereditary conditions of late manifestation, and to what extent with end results of social and environmental influences that have impinged during the life span? Again, the aged have been called the first poverty-stricken leisured class in history. How far can the emotional disturbances, now discovered to be relatively widespread among them, be attributed to the many adverse pressures emanating from their present environment, their poverty, isolation and lack of prestige? Ageing is certainly out of fashion and unpopular in some cultures, and this is considered by some observers to be the heart of the matter of mental health and adjustment in ageing. Following from this, how far could we expect the incidence of psychiatric disorders of old age to respond to preventive measures aimed at relieving their social and economic hardships? Inquiries in recent decades have made it possible to suggest tentative answers to some of these questions. In describing our own inquiries, I shall refer to the work carried out by a group of people who included Drs Kay, Garside, Beamish, Bergmann, Blessed and Tomlinson, who have kindly allowed me to refer to a number of inquiries we have undertaken as a team.

VARIETIES OF MENTAL DISORDER IN OLD AGE AND THEIR AETIOLOGICAL BASIS

Mental disorder in old age readily evokes the image of bleak dormitories in institutions full of helpless old people suffering from senile mental decay. Such patients are indeed to be found in mental hospitals, residential homes and hospitals for the chronic sick. However, in psychiatry as in medicine, inquiries in recent decades have shown that much that was formerly attributed to senility was often due to some specific disease that occurs also in early life and is merely given a special colouring by senescence. The demonstration that many of these disorders could be successfully treated has injected a spirit of optimism into geriatric psychiatry, although the progress made is of course dwarfed by the ignorance that remains.

It has been known for a considerable time that mental disorder in old age may present in a number of different ways with depressive, manic, paranoid, delirious or neurotic symptoms, or with a gradual and progressive decline of memory and intellectual functioning. But the part played by cerebral degeneration and other forms of physical disease in the causation of some of these syndromes was left in uncertainty by the older descriptions. Some of the reasons for this uncertainty have become more plain in recent years. In the first place, degenerative changes of a kind first described in association with senile dementia half a century ago, are now known to occur commonly in the brains of normal old people. A second factor that may have contributed is the frequency in old age of minor neurological signs of mild memory defect and of physical disorders such as hypertension, cardiac disease and general atherosclerosis, whose contributions to the causation of an associated psychiatric syndrome are difficult to assess in the individual case. Some two decades ago, attention began to be drawn to the favourable response often made by patients showing depressive or paranoid symptoms to electroconvulsive therapy.[4, 17] A number of follow-up studies [3, 20, 21] also showed striking differences in outcome between groups with predominantly ' functional ' and predominantly ' organic ' symptoms.

The hypothesis that there existed among the mental disorders of old age a number of distinct nosological entities in addition to senile and arteriosclerotic dementia was tested some fifteen years ago by subdividing patients according to the psychiatric symptoms and signs they presented alone and observing their outcome after a period of time; with the exception of evidence indicative of an unequivocal cerebrovascular infarction, all physical illness and disability present was ignored in making a diagnosis. It was found that the tentative groupings of affective disorders, paraphrenia of late onset, acute and sub-acute delirious states and senile and arteriosclerotic psychoses behaved quite differently at six months, two years and at seven or eight years after admission to hospital.[22, 24] The mortality in the last two groups and the discharge rate in the first three, differed in a highly significant manner and the difference was to a considerable extent independent of age. The differences between the groups were confirmed by the results of formal, psychological testing [7, 23] and by post-mortem findings in the different groups.[9] Moreover, detailed follow-up inquiries have shown that in cases tentatively diagnosed as 'functional' (unrelated aetiologically to cerebral disease) the decay of intellect and personality characteristic of dementia is exceptional.[8, 19, 22]

Comparisons with the expected mortality in the general population of corresponding age have shown that the late schizophrenic and paranoid disorders have a normal expectation of life, while patients with depressive or manic symptoms survive for about seven-tenths of the normal life span. On the other hand, where a senile or arteriosclerotic dementing process had been initially diagnosed, length of survival has been found to be about one-quarter of the normal, mortality being at least five times as high as in the general population.[8] An interesting paradox emerged here, which may have biological significance. In all the clinical groups studied a proportion of the deaths had been attributed in case records and on the death certificates to vague causes, such as myocardial degeneration, 'old age', or bronchopneumonia. But in the organic groups, i.e. in those with the highest mortality, the death rate from such cases was eight times as high as that recorded in the general population, and in cases of senile dementia nearly all the deaths were attributed to such causes. On the other hand, the proportion of such non-specific deaths did not depart from the normal rate in the affective and paraphrenic groups, and these had the lowest mortality. Degenerative disease in the brain, therefore, appears to cause death by undermining resistance to minor infections and insults—in other words, in a manner that resembles senescence itself.

HEREDITY AND ENVIRONMENT

To what extent are we dealing in these disorders with the effects of hereditary predisposition on the one hand, and social and environmental factors on the other ?   As far as the senile and arteriosclerotic psychoses are concerned, although a good deal is known about their underlying pathology, the responsible causes are obscure.   Genetic factors probably contribute in causation but they are almost certainly not of the classical, single monohybrid kind.   Among the delirious cases, the great majority—some 85 per cent—owe their condition to a specific physical disease;   and this fact is of great importance from a practical point of view, because unless a reliable history is available the most severely ill and confused of these subjects are prone to be misidentified as suffering from irreversible cerebral degeneration.

As far as the affective and schizophrenic disorders of late life are concerned, the analogous forms of illness in earlier life are known to be determined to some extent by hereditary factors of a specific kind.   It is of considerable interest that these genetical factors have been shown to make a significantly smaller contribution to the causation of the senescent forms of these disorders.   Thus in cases of affective disorder of late onset, the morbid risk for these conditions among the near relatives is approximately 4·5 per cent, which is higher than that expected for the population at large, but significantly lower than among cases of early onset.[8, 26]   Again, in the case of the schizophrenias of late life, the morbid risk for this disease among first-degree relatives is only slightly above that for the normal population, and substantially below what has been recorded in most authoritative studies of schizophrenia in earlier life. These findings are in accord with expectations on general biological grounds ; one would anticipate that those most markedly predisposed to breakdown would manifest their propensity at a comparatively early stage of the life span, and also that in those most resistant genetically speaking, the development of illness would be postponed till an advanced age.

It would be further anticipated that in subjects in whom both low genetic loading and late onset of illness reflect a relatively stable constitution, stressful circumstances would contribute in a more conspicuous manner than in those more strongly predisposed by heredity.   Investigation has, in fact, confirmed [8] that chronic, progressive and disabling physical

M

illness contributes to the causation of the depressions of late life, and that bereavement, economic hardship and other stressful circumstances play a part in the precipitation of illness in three-quarters of cases of depression of late onset, as compared with half of those who had also suffered an illness in earlier life. A similar situation was found in the late schizophrenias; deafness, cerebral disease and, to a very limited extent, social isolation all contribute to causation in a proportion of cases.[10]

THE ASSOCIATION BETWEEN CEREBRAL DEGENERATION AND MEASURES OF PSYCHOLOGICAL DETERIORATION

I have already referred to the ubiquity of some degree of degenerative change in the brain and the uncertainties this has engendered concerning the causation of mental disorder in the aged. In the absence of any technique of measurement, these changes have, in the sixty years that have elapsed since they were discovered, been alternately regarded at some times as explaining everything and at other times as being largely irrelevant and explaining little or nothing. More recently it has been demonstrated that the ubiquity of the changes may be due to their association with a quantitatively graded effect, a change that to some extent cuts across diagnostic distinctions; I mean the gradual decay in certain aspects of intellectual and personality functioning that is in some measure universal in old age. Quantitative measures of intellectual performance and behaviour in everyday tasks were recently made in a large number of patients with and without psychiatric symptoms who had been admitted to psychiatric, geriatric and general hospital wards; and the relationship of these measures to quantitative estimates of the severity of senile plaque formation in the cerebral cortex was investigated in those subjects on whom post-mortem examinations could be conducted. Senile plaques are discrete clumps of granules or fibres in the cerebral grey matter, which show up with silver stains. They can be counted in sections of cerebral cortex and with the aid of a sampling technique, a mean count for the subject can be computed.

Mean plaque counts have been found to have a highly

significant correlation with measures of proficiency in every-day practical tasks, and of intellectual functioning, carried out during life.[2, 25]   It is of considerable interest, however, that the effects exerted by processes of which the accumulation of plaques is but one index, appear to be governed by a threshhold phenomenon.   Thus, subjects with a mean count of less than 5 are rarely demented clinically, whereas the majority of those with mean counts of more than 7 show obvious mental deterioration.   It would seem possible therefore that fall-out of cells may be compensated, or damage accommodated, within the reserve capacity of the cerebral neurones until it exceeds a certain level.   If these observations can be confirmed, they will provide evidence for the view that the difference between the mentally well-preserved old person and the senile dement suffering from the commonest form of cerebral degeneration in old age arises from a differing rate of progression of the same process.

PSYCHIATRIC DISORDERS OF OLD AGE IN THE COMMUNITY

The clinical studies of affective neurotic and paranoid disorders directed attention to exogenous and environmental factors that may contribute in causation.   These cannot be studied in a satisfactory manner in hospital populations, since adverse environmental factors could well have had a selective influence on admission.   To investigate the part played by social factors in the causation of mental illness in the elderly, we have therefore to turn our attention to community samples. For the community, one of the most important facts about psychiatric disorders of the aged is the enormous size of the problem as shown by findings in a number of surveys.[5, 11, 12, 13, 18, 30]   In the Newcastle Survey, for example, the prevalence of severe organic psychosis was 5 per cent and the total prevalence, including mild cases, reached 10 per cent of the population aged sixty-five and over.   The prevalence of major functional psychoses was 2·4 per cent, most of these cases being moved to hospital at a fairly early stage of their illness.   Moreover, 10 per cent of subjects were regarded as suffering from a moderate to severe neurotic disturbance (usually presenting as a blend of depression and anxiety), while a mild neurotic illness was judged to be present in a further 14 per cent of subjects. The total prevalence of all forms of functional disorder was 31 per cent. In approximately half the organic cases ascertained in the community the deterioration was similar in degree to that found in demented hospital patients and fewer than one-fifth of the total in this group were in fact being treated in hospital or a residential institution. In our own study, nearly three times as many people aged sixty-five and over, with relatively severe psychiatric illness, were living in the community as there were in all forms of hospital and residential accommodation; and only a small fraction of them were receiving any form of psychiatric help.

When these facts are coupled with the finding that some 4·5 per cent of people aged sixty-five and over in institutions are enough to cause considerable strain on the resources of the Health and Welfare Services, it becomes obvious that there is no alternative to a community-based service offering a wide range of facilities. Long-term or permanent institutionalization of old people must become a last resort applied only in rare instances.

A community service, to be effective, would have to direct special attention to individuals known to be vulnerable to breakdown. In an attempt to define indices by means of which such vulnerable old people could be identified, we have undertaken statistical analysis of a large number of items relating to psychiatric and physical illness, personality characteristics, social class, income and social isolation and loneliness. Among the indices which characterized those found to be suffering from a functional or organic illness, were such features as advanced age, fewness of everyday contacts, decreasing frequency of social visiting, impaired mobility and capacity for self-care, a history of previous psychiatric illness, a limited range of interests in the past, low sociability and marked tendencies in the premorbid personality to moodiness, anxiety or hypochondriasis. A number of features relating to physical health, such as the presence of a definite physical disability, a feeling of being in poor health, liability to falls, impairment of hearing and vision and a physical appearance suggesting ageing beyond the individual's years, were all highly correlated with psychiatric disability; and it was interesting and important that the association was quite marked in the functional as well as in the organic groups of psychiatric disorder. High income showed negative correlation with illness, but social class had no association with functional or organic mental disease. Complaints of loneliness, attitudes of self-pity and of dissatisfaction with life, as also general feelings of malaise, proved to be highly correlated with a diagnosis of psychiatric disorder and of functional illness in particular, and so were symptoms of anxiety and depression. To state this may appear to be labouring the obvious, but it is perhaps of some importance from a preventive point of view that, in a follow-up study of our community sample four years after the initial ascertainment, a high correlation was found between the presence of isolated symptoms of anxiety or depression not severe enough to warrant a definite psychiatric diagnosis, and the finding of definite illness on the second visit.[1] Again, as was to be expected, symptoms such as impairment of memory, disorientation and clouding of consciousness were correlated with a diagnosis of organic mental disease. But from the point of view of ascertainment and prediction, the association between minor symptoms of this nature discovered in the initial interview and a relatively high mortality in the intervening period, and also the finding of dementia in a substantial proportion of these subjects on follow-up, were important and unexpected.[14]

THE ASSOCIATION BETWEEN SOCIAL VARIABLES AND MENTAL ILL-HEALTH

It is obvious that features such as those listed could prove valuable for any service concerned with the ascertainment of mental ill-health at a relatively early stage in order to make a selective deployment of scant resources. However, these correlations in themselves neither tell us very much about the interrelationship between the different features, nor permit us to begin the process of disentangling cause and effect. To do this and, in particular, to make possible a clearer appraisal of the role of social factors, the intercorrelation of 34 features relating to social status, physical and mental health, personality and predisposition in 267 subjects was calculated, and a principal component analysis carried out with the aid of a Pegasus computer.[5]

Only a few of the findings will be selected for comment here. Income had a negative correlation with psychiatric illness, but social class bore no significant relationship, that is, neither functional nor organic illness showed a differential class incidence. Surprisingly, neither did bereavement show any significant correlation with mental illness. This may have been due to a too comprehensive definition of the term, although a high proportion of those described as 'bereaved' had suffered widowhood. Again, although illness had tended to decrease the number of human contacts made by the patient, isolation as a 'factor' derived from the intercorrelation of a large number of measures, proved to have very little association with illness. In a 'surviving' population such as that studied, isolation appeared to be characteristic of two types of individual. The first were the exceptionally hardy, who adjusted themselves successfully and were able to live alone and care for themselves in old age, despite lack of relations. The others were life-long isolates for whom being alone had become a mode of life. It is self-evident that illness will tend to isolate individuals to some extent, but it is to be doubted whether, as is sometimes assumed, isolation is always harmful in its effects or an indication of serious personality deviation. This subject has been extensively investigated in the United States by Lowenthal [16] and her conclusions about the role of isolation are along similar lines.

The possibility has to be considered that those negative findings merely reflected the fact that we were dealing with a 'survivor' population from which the socially underprivileged, the bereaved and the isolated sick had been removed to institutions. The evidence suggested that to a limited extent such was indeed the explanation. Thus there was a marked over-representation of individuals drawn from Social Class V in the geriatric wards and welfare homes serving the areas concerned.[11] Moreover, in geriatric medical wards and welfare homes we found a striking preponderance of the single, widowed and divorced, and it seemed clear that many of them had lived in rather isolated conditions before admission. As a third of the geriatric patients were suffering from organic psychoses, minor degrees of deterioration or functional disturbances, it was plain that such institutions draw away from the community a substantial number of isolated and psychiatrically disturbed old people. Poverty and adverse circumstances might also have acted by removing individuals from the population by death at an earlier age. Even if the only medical effect of social privation is to make institutionalization more likely, this is important. It is perhaps relevant in this connection that more than half a million people were entitled to have their pension supplemented by the National Assistance Board, but did not apply either because they did not know they were so entitled or because for them the Poor Law stigma still lingered.[29]

However, if we return to the contribution of existing social stresses to the causation of actual illness, the findings in this community study suggest that this could only have been a limited one. For the community sample contained the overwhelming majority of cases in every category of disorder, missing only the very small group of late schizophrenias. There were six times as many organic syndromes, twelve times as many functional psychoses and forty times as many neurotic disorders in the community as there were in institutions. Even if current social factors are allowed far greater importance in relation to the institutionalized cases, inclusion of them along with the community sample would not have altered the overall picture, as far as relationship between social factors and illness was concerned.

As far as relieving the effects of isolation is concerned, it was of interest that assessments of the degree of loneliness proved to be highly correlated with lifelong tendencies to anxiety, moodiness, lack of sociability, history of previous illness and symptoms of anxiety and depression rather than with quasiphysical terms as measured by such indices as ' living alone ', ' frequency of visits ' and ' frequency of contact with relatives ', with which they proved to have little association. In other

words, complaints of loneliness are generally symptomatic of
life-long personality traits and of the psychiatric disorders they
are apt to engender in old age. Loneliness is a feeling of
neglect, expressed by individuals with certain types of psychiatric illness. It is unlikely to respond, unless attention is
directed towards removal of the underlying causes rather than
to the improvement of social contacts.

As far as the possibilities of effective prevention (which are
to be discussed in the next section) are concerned, the highly
significant association between a diagnosis of functional,
psychiatric disorder and certain indices of physical ill-health,
may prove to be important. The commonest disorders found
to be associated were cardiovascular and respiratory disease
and joint, muscle and connective disease. The association was
particularly striking in males. A follow-up inquiry,[14] carried
out four years after the initial ascertainment, showed that these
individuals had died at a significantly greater rate than those
without psychiatric disorder, or those in the general population
of comparable age. The disabilities complained of and the
feelings of ill-health described could not therefore have been
to any appreciable extent the expression of hypochondriacal or
depressive symptoms. However, the physical and psychiatric
disabilities, although correlated, showed in their progression
considerable independence of each other. There are, therefore,
strong indications for trying to identify such patients and
treating their psychiatric disorders at an early stage. Their
emotional distress can usually be relieved, although at the
present time help is only rarely being provided. Whether
such treatment influences the mortality rate of this group of
subjects or not is a matter for future inquiries.

THE POSSIBILITIES OF PREVENTION

These facts suggest that only in a very restricted sense can
we speak hopefully or realistically about preventing mental
disorder among the elderly. The most immediate and pressing
need is for measures that would help to bring sufferers from
psychiatric illness to notice at an earlier stage in the development of the disability. Only a small fraction of those requiring

help are receiving it; the amount of distress, and perhaps of chronic emotional disorder, might well be reduced by earlier diagnosis and treatment. Measures directed to increasing pensions, to preparing for retirement, to improving housing conditions that are often deplorable and to relieving loneliness are desirable and urgently needed. Moreover, they are justified in their own right, to relieve existing suffering and bring dignity and fullness to the life of the old person; many old people complained of hardship or loneliness without being ill. Nevertheless, even if we allow for the fact that community samples are a ' surviving' population, it is highly improbable that the measures listed or those which could be directed towards the relief of social isolation or poverty, would substantially reduce the incidence of mental disorder among the aged.

However, the benefits that would flow from such general social measures would be far from negligible. In Great Britain, the unmarried, the widowed, those without children and brothers and sisters are markedly over-represented in institutional accommodation. Moreover, even if action designed to break down isolation does no more than help more old people to remain in their own homes, this not only accords with their wishes, but is of vital importance for Health and Welfare Services. At the present time, inquiries in a number of countries have shown that 80 per cent of old people enjoy close and continuing contact with their relatives, often with reciprocal flow of services, important for young and old alike.[27, 28] Moreover, it has been estimated that if it were not for the aid given by relations, the pressure on the Health and Welfare Services would be four or five times as great as it is.[27] Even if part of this burden were to be relinquished, within a short space of time serious difficulties would arise. For the 4·5 per cent of people aged sixty-five and over, who are resident in some kind of institution, are enough to cause serious overcrowding in many places and long waiting-lists. We cannot assume, however, that such family cohesion will continue indefinitely, for it is almost certainly being undermined by social and geographical mobility. Unless imaginative counter measures are adopted, these factors, together with the brick and mortar planning of new communities, may slowly erode the means whereby families have cared for their elderly members for generations.

ENVIRONMENTAL FACTORS IN HISTORICAL PERSPECTIVE

Social factors, then, had emerged from the cross-sectional analysis of the characteristics of elderly people with a very modest role, as far as the prospects of prevention of actual illness in the short-term were concerned.   But it has to be asked whether in the psychiatrically ill patients we could have been observing in part the end results of social and environmental factors that had first impinged in the remote past. The factors involved could well remain hidden from a cross-sectional investigation of presenting features but emerge more clearly from a longitudinal or historical analysis.   Such information as was elicited about the life histories of the patients with affective disorder suggested that there had very likely been some such contribution towards the end result.   It is of course very difficult to disentangle the effects of inherent predisposition from the causes of adversity and maladjustment, but the attempt has to be made.

The following are some of the relevant findings.[13] Anomalies of personality had often been consistent and lifelong, and some contribution from genetic factors may be presumed.   However, a significantly higher proportion of patients with affective disorder had lost a parent during childhood than had normal subjects, and tendencies to anxiety and moodiness had been present from a relatively early age.   The higher frequency of divorce and separation, the smaller number of children born within the marriages of male patients, the more restricted social life and narrowness of interests, and the earlier attacks of illness in some subjects, may have been expressions of the same inherent predisposition.   But the events in earlier life and what is known of the aetiology of neuroses suggest that we were almost certainly observing the results of an interaction between social, familial and genetic factors and not the effects of heredity alone.

In any event, these vicissitudes had further limited the

chances of successful adjustment at the next stage of the life span.   When we arrive at old age we find in this functional group, as compared with normals, earlier retirement and fewer contacts.   They are more often living alone and have poorer amenities in the home.   The later stresses of physical illness, bereavement and further reduction in human contacts are then impinging upon vulnerable personalities.   The view from the ' wrong end of the telescope ' looks in some ways like a protracted game of chess in which one side has had some of the more powerful chessmen eliminated at an early stage of the game, restricting mobility and manœuvring and progressively reducing the chances of a successful end to the game.

In other words, as far as the commonest forms of emotional disturbance in old age are concerned, the lesson taught by extended history-taking is similar to that which emerges from every field of inquiry, whether it concerns the physical health, employment, adjustment, housing or leisure of the aged.   It is that, if we wish to intervene effectively, we must take action long before senescence.   We have yet to learn how to achieve this.

SUMMARY

1. Although cerebral degeneration is responsible for the most intractible forms of psychiatric disorder in old age, the greater part of the morbidity found in the community and more than half of the illnesses admitted to psychiatric hospitals in senescence, are unconnected with degenerative change. Most old patients suffer from conditions that can be mitigated, although help is rarely being provided at the present time.

2. Among the functional mental disorders hereditary factors appear less important than in the comparable conditions of earlier life.   Environmental and exogenous factors are more important.

3. The recent demonstration of a quantitative and orderly relationship between measures of the commonest forms of pathological change in the brain in old age and estimates of psychological functioning may provide the basis for fresh inquiries into certain problems, particularly those in the borderlands of normal and pathological mental ageing.

4. Social class, isolation and bereavement exert an important influence on chances of admission to hospital, and may contribute to causation of illness. But when the role of these factors is studied in a representative sample of the population, their contribution is found to be limited. Although they indubitably cause hardship, they probably give rise to actual illness mainly in those with a pre-existing vulnerability.

5. Social and environmental factors probably contribute to causation in less tangible ways in earlier stages of the life span. There is evidence to suggest that in some individuals predisposed to functional disorder, constitutional and socio-familial factors limit the possibilities of successful adjustment in a cumulative manner. The process culminates in breakdown in a proportion of subjects.

6. Among the indices which differentiate the psychiatrically ill from normal elderly subjects are such features as advanced age, few everyday contacts, impaired mobility, history of previous psychiatric illness, limited interests, low sociability and tendencies to moodiness and anxiety in the pre-morbid personality. There is also a close relationship between physical illness and functional psychiatric disorder in old age. Such features may prove valuable guides to devising programmes of ascertainment and preventive care.

7. Measures to remedy poverty and isolation in the aged are required in their own right, and early treatment for those suffering from psychiatric disorder and the commonly associated physical illnesses is an urgent need. In the light of the inquiries discussed here, however, it is unlikely that such measures will exert much influence upon the incidence of mental disorder in the elderly.

DISCUSSION

In reply to a questioner PROFESSOR ROTH said that his findings accorded with certain predictions that had been made, although he wondered whether these had been based on valid assumptions. He and his co-workers had predicted that in both the affective and schizophrenic cases, those who suffered their first attack in late life would show a lower genetic loading than those with a first attack in early life. Correspondingly they had expected that the cases of late onset would be found to have suffered more environmental stress prior to breakdown than the cases with an early age of onset. However, this situation is perhaps not simple enough for the assumptions made to be necessarily valid. For example, if the hereditary basis of all schizophrenias is a major gene, the morbidity risk among the first-degree relations of schizophrenics perhaps should not significantly differ from the corresponding risk in early life cases. However, this raised a controversial though interesting issue which led away from the main theme.

The schizophrenias of late life, Professor Roth continued, provided a good example of hereditary factors potentiating the effects exerted by an adverse environment. As measured by a number of indices, they proved to be rather more isolated than in comparable subjects suffering from an affective disorder. He and his co-workers had attempted, therefore, to investigate whether isolation could have contributed to the causation of their illness, or whether it merely reflected the deviations of premorbid personality that tend to be common in schizophrenics. The answers they had obtained to the questions raised proved complicated and interesting. The schizophrenics were found to have had far more abnormal personalities than the affective cases, and the traits they had manifested would have been expected to render social contacts difficult. They were frequently described in terms such as arrogant, argumentative, quarrelsome without cause, cold-hearted, sensitive, jealous, eccentric. Thus schizoid or paranoid traits were common, and it was probably for this reason that the unmarried had been over-represented in this group. An elegant piece of evidence relating to these premorbid personality traits had been elicited from a comparison of the illegitimacy rates in the late schizophrenic and the late affective groups. Among a group of cases studied by Dr Kay in Stockholm, more than one-quarter of the forty-eight female patients had had illegitimate children. But, whereas almost everyone of the cases among the affective group in a similar predicament had married the father of the child, only one of the schizophrenics had subsequently done so. Hence, almost half a century before the majority of schizophrenic subjects had come under psychiatric observation, their personalities were schizoid in a distinctive manner. The abnormalities had, in other words, been lifelong and could probably be attributed to a large extent to genetic causes.

However, some of the factors which had added to the isolation of these individuals in old age had clearly been independent of their lifelong personality traits. Thus they were significantly more frequently drawn from the younger members of sibships and had, in fact, fewer relatives alive. Those that had married had had fewer children, and a smaller proportion of these were alive than in control groups. Finally, deafness coming on in late life, which could be expected to enhance their isolation, had been significantly more common than in control groups. The conclusion had been that these individuals had to a large extent been self-isolated and that their marginal adjustment over the greater part of life had probably been due to genetically determined personality traits. However, it was not until accidental factors in old age pressed their isolation beyond a certain threshold that actual breakdown had occurred. Hence this was a good example of hereditary and environmental factors potentiating each others' effects.

REFERENCES

1. BERGMANN, K. 1966. Unpublished observations.
2. BLESSED, G., TOMLINSON, B. E. and ROTH, M. 1967. In the press.
3. CLOW, H. E. 1948. Outlook for patients admitted to mental hospital after the age of sixty. *N.T.S. J. Med.* **48**, 2357.
4. GALLINEK, A. 1948. Nature of affective and paranoid disorders during senium in light of electric convulsive therapy. *J. nerv. ment. Dis.* **108**, 293.
5. GARSIDE, R. F., KAY, D. W. K. and ROTH, M. 1965. Old age mental disorders in Newcastle-upon-Tyne. III. A factorial study of medical, psychiatric and social characteristics. *Brit. J. Psychiat.* **111**, 939.
6. GOLDHAMER, H. and MARSHALL, A. W. 1953. *Psychosis and Civilization.* Glencoe, Illinois.
7. HOPKINS, B. and ROTH, M. 1953. Psychological test performance in patients over sixty. II. Paraphrenia, arteriosclerotic psychosis and acute confusion. *J. ment. Sci.* **99**, 451.
8. KAY, D. W. K. 1959. Observations on the natural history and genetics

of old age psychoses : a Stockholm material, 1931–1937. (Abridged.) *Proc. R. Soc. Med.* **52**, 791.

9. KAY, D. W. K. and ROTH, M. 1955. Physical accompaniments of mental disorder in old age. *Lancet* ii, 740.

10. KAY, D. W. K. and ROTH, M. 1961. Environmental and hereditary factors in the schizophrenias of old age ('late paraphrenia') and their bearing on the general problem of causation of schizophrenia. *J. ment. Sci.* **107**, 649.

11. KAY, D. W. K., BEAMISH, P. and ROTH, M. 1962. Some medical and social characteristics of elderly people under state care. *The Sociology Review Monograph*, No. 5. Keele. p. 173.

12. KAY, D. W. K., BEAMISH, P. and ROTH, M. 1964a. Old age mental disorders in Newcastle-upon-Tyne. I. A study of prevalence. *Brit. J. Psychiat.* **110**, 146.

13. KAY, D. W. K., BEAMISH, P. and ROTH, M. 1964b. Old age mental disorders in Newcastle-upon-Tyne. II. A study of possible social and medical causes. *Brit. J. Psychiat.* **110**, 668.

14. KAY, D. W. K. and BERGMANN, K. 1966. Physical disability and mental health in old age. *J. psychosom. Res.* **10**, 3.

15. KAY, D. W. K., BERGMANN, K. and GARSIDE, R. F. 1967. A four-year follow-up study of a random sample of old people originally seen in their own homes. A physical, social and psychiatric enquiry. *Communications to the 4th International Congress of Psychiatry, Madrid.* In Press.

16. LOWENTHAL, M. F. 1963. Some social dimensions of psychiatric disorders in old age. In *Processes of Ageing*, Vol. II. (Ed. R. H. Williams, C. Tibbitts, W. Donahue.) New York. Atherton.

17. MAYER-GROSS, W. 1945. Electric convulsive treatment in patients over sixty. *J. ment. Sci.* **91**, 101.

18. PARSONS, P. L. 1965. Mental health of Swansea's old folk. *Brit. J. prev. soc. Med.* **19**, 43.

19. POST, F. 1951. The outcome of mental breakdown in old age. *Brit. med. J.* i, 436.

20. POST, F. 1962. *The Significance of Affective Symptoms in Old Age.* Maudsley Monograph, 10. Oxford Univ. Press.

21. ROBERTSON, E. E. and BROWNE, N. M. 1953. Review of mental illness in the old age group. *Brit. med. J.* ii, 1076.

22. ROTH, M. 1955. The natural history of mental disorders in old age. *J. ment. Sci.* **101**, 281.

23. ROTH, M. and HOPKINS, B. 1953. Psychological test performances in patients over sixty. I. Senile psychosis and the affective disorders of old age. *J. ment. Sci.* **99**, 439.

24. ROTH, M. and MORRISSEY, J. D. 1952. Problems in the diagnosis and classification of mental disorder in old age. *J. ment. Sci.* **98**, 66.

25. ROTH, M., TOMLINSON, B. E. and BLESSED, G. 1966. Correlation between scores for dementia and counts of 'senile plaques' in cerebral grey matter of elderly subjects. *Nature* **209**, 109.

26. STENSTEDT, A.   1959.   Involutional melancholia, *Acta psychiat. scand.* Suppl. 127.

27. TOWNSEND, P.   1957.   *The Family Life of Old People.*   London.   Routledge & Kegan Paul.

28. TOWNSEND, P.   1962.   *The Last Refuge:* a Survey of Residential Institutions and Homes for the Aged in England and Wales. London.

29. TOWNSEND, P. and WEDDERBURN, D.   1965.   *The Aged in the Welfare State.*   Occasional Papers on Social Administration No. 14.   London. Routledge and Kegan Paul.

30. WILLIAMSON, J. *et al.*   1964.   Old people at home : their unreported needs.   *Lancet* i, 1117.

AGEING AND MORTALITY

R. D. CLARKE

Continuous Mortality Investigation Bureau,
Institute and Faculty of Actuaries

THE study of death rates is a branch of scientific inquiry which has been pursued for more than three centuries. The pioneer in this field was an Englishman, John Graunt, who in 1662 published a study based upon the Bills of Mortality recorded by the authorities of the City of London.[5] Some thirty years later the astronomer Halley, of comet fame, constructed a table of mortality from the records of deaths in the city of Breslau.

It is possible to claim an even greater antiquity for the mathematical study of mortality, since a rudimentary life table was constructed by Ulpian during the time of the Roman Empire and was used for the purpose of calculating annuities. It had a long history and was still in use in Italy some fifteen centuries later. However, there is no evidence that this table was based upon statistical observation and it is thought likely that Ulpian invented it as a piece of intelligent guesswork. It cannot therefore be regarded as belonging to the history of scientific demography.

Nowadays tables of mortality are usually constructed from statistical data—I say 'usually' because it is quite possible to construct a hypothetical table without any statistical foundation at all. Even when the rates of mortality have been observed statistically, there is a sense in which a mortality table is an artificial construct. The observed death rates, which form the basis of such a table, normally relate to a limited period. Thus the English Life Table No. 11 is based upon deaths occurring in the three years 1950–2. But the mortality table itself purports to follow a group of new-born infants throughout life and it is highly improbable that children born in 1950–2 will themselves be subject to the rates of mortality experienced by the general population at that tine. For this reason, therefore,

N   183

English Life Table No. 11, like most of the standard tables known to actuaries and demographers, is essentially a hypothetical model.

It is, of course, possible to construct genuine life tables for past generations. Thus, for the generation born in 1850, the national records enable rates of mortality to be calculated for each year up to the present time and so a generation life table or, as it is sometimes called, a cohort life table can be constructed. A whole series of such cohort tables has in fact been produced by Dr. Case [1] of the Institute of Cancer for use in the statistical analysis of cancer mortality. In general, however, the drawback of cohort or generation tables is that they cannot be completed until the generation concerned is extinct; and since we want to keep up to date in our knowledge of human mortality, we have to construct tables based upon recent experience, even though we recognize the artificiality of the model we are using.

The reason why we have to keep up to date is that for at least a century, and possibly for much longer, rates of mortality have been falling in this country. Indeed, at the present time mortality rates are falling all over the world and this fact is a major cause of what has come to be known as the ' population explosion '. Forecasting the future trend of mortality—admittedly a hazardous undertaking—has become a vital concomitant in estimates of future population figures.

As everyone knows, the rate of mortality varies from age to age. As Professor McKeown has shown it is desirable for many purposes to divide the first year of life further, and to compute the mortality rate for the first week of life, or the first four weeks, and so on. However, infant mortality is a topic which, important as it is, does not concern us in a discussion of the relation between mortality and ageing, and I shall therefore confine the present discussion to mortality from the age of twenty onwards.

Broadly speaking, the rate of mortality in adult life increases with age. However, this is not universally true; for example, motor accidents at the present time take a particularly heavy toll among young men, and as a result it is not unknown for investigations to show a higher death rate for men at age twenty than at age thirty. In former times, tuberculosis reached a peak in both sexes in the late twenties, so that the mortality would rise to a maximum round about age twenty-eight and then fall again before resuming an upward trend. In former times there used to be also an appreciable mortality risk associated with childbearing, with consequent distortion in the female mortality curve over the childbearing range of ages.

With these exceptions in mind, however, we can return to the general proposition that in adult life the rate of mortality increases with increasing age. It may be helpful to illustrate this with some rates of mortality for the general population of England and Wales, taken from English Life Table No. 11,[9] which reflects mortality in the three years 1950–2. For adult men the rate of mortality was 0·00129 at age twenty, increasing to 0·00290 at age forty, 0·02369 at age sixty, 0·13629 at age eighty and 0·44045 at age one hundred. Thus in the twenty-year interval between ages twenty and forty the rate of mortality approximately doubled; in the next twenty-year interval it multiplied by a factor of 8·2; in the third interval (between sixty and eighty) it further multiplied by a factor of 5·75; and in the fourth interval it multiplied by a factor of just over 3. This means that the steepest increase lay between forty and sixty, the range of ages where, for men, lung cancer and coronary disease begin to take their toll.

The corresponding mortality rates for women are substantially lower than for men. At age twenty the rate is 0·00083, increasing to 0·00227 at forty, to 0·01271 at sixty, to 0·10466 at eighty and 0·36764 at one hundred, but the interesting feature of these rates is that the twenty-year increase factor in the two middle intervals shows a reverse situation to that of the men's experience. Between forty and sixty the increase factor for women is 5·6 as compared with 8·2 for the men, while between sixty and eighty it is 8·2, as compared with 5·75 for the men. Thus the steepest increase for women lay between ages sixty and eighty. The interpretation of this contrast between the two sexes lies in the fact that there is a smaller impact on female mortality of lethal diseases during the period of middle age. Consequently the curve for female mortality remains at a lower level than the male curve and then has to take a sharper upward swing when passing from middle age into old age. Even so, the female mortality curve never merges with the male curve and, broadly speaking, lags five years behind it. In fact we can say that in England and Wales a woman, considered as a mortality risk, is roughly equivalent to a man five years younger than herself. Whether one can pursue this conclusion a stage further and say that the ageing process in woman lags five years behind the corresponding process in man is a question which, in rather more general terms, I shall attempt to discuss later.

The general tendency for the rate of mortality to increase throughout life has led theorists to speculate on the possible existence of what is commonly termed a 'law of mortality', i.e. a mathematical formula which can relate the rate of mortality to the attained age. The earliest of such speculations was that of Benjamin Gompertz,[4] who in 1825 expressed a law of mortality by the following formula:

$$\mu_x = Bc^x$$

in which x is the attained age and μ_x is the force of mortality at that age. The 'force of mortality' is very similar to the rate of mortality except that at the upper limit it tends to infinity and not to unity or some other finite quantity. Obviously, so long as the parameter c is greater than unity, the term Bc^x is also going to tend to infinity.

Gompertz gave his formula a philosophical rationale. He assumed that man's 'inability to withstand destruction', as he termed it, increased in equal proportion over equal successive intervals of age. The chief interest of this approach to-day is that it shows that Gompertz was concerned to link the statistically observed rate of mortality with some force or process inside the human organism. A declining ability to resist destruction may not be exactly identical with the concept of ageing which we are considering at this Symposium, but I think it is sufficiently close to indicate that Gompertz visualized a direct link between a measurement of the ageing process and the observed rate of mortality.

Gompertz's law of mortality has been modified by later writers. In 1889 William Makeham[6] added a term to cover deaths caused by chance events, so that the formula became :

$$\mu_x = A + Bc^x$$

Then in 1931, W. Perks [8] proposed two further modifications. His first formula was :

$$\mu_x = (A + Bc^x)/(1 + Dc^x),$$

which can also be written :

$$\mu_x = K - 1/(\alpha + \beta c^x).$$

This is a sigmoid curve of a type known as a logistic and it allows for the fact that in extreme old age there is a tendency for the rate of mortality to taper off instead of continuing to increase at an accelerating rate. Perks's second formula was :

$$\mu_x = (A + Bc^x)/(Kc^{-x} + 1 + Dc^x).$$

This has five parameters and is considerably removed from the simplicity of Gompertz's original law ; and more parameters still have been required in more recent formulas. Thus in 1956 seven parameters were found necessary to represent the adult section of the mortality curves for English Life Table No. 11. Such formulas make it virtually impossible to evolve a philosophical rationale connecting the observed rate of mortality with such concepts as the ageing process or the inability to withstand destruction. In other words, although I believe that there is a real connection between mortality and ageing, I am at the same time convinced that attempts to demonstrate this relationship in terms of a tidy mathematical formula are doomed to failure. The main purpose of using such a formula nowadays is to produce a smooth curve to replace the inevitable unevenness of the curve of observed data. Technically this is known as the process of ' graduation '.

Now let us turn from the rate of mortality and consider a different function, namely the d_x column of the mortality table, or ' curve of deaths ' as it sometimes called. As before, I ignore the ages below twenty so that we have a statistical distribution with a shape that is fairly familiar, except that it is markedly asymmetrical, extending very much farther below than above the maximum value, or mode. In the English Life Table No. 11 for men it begins at the low level of 123 deaths at age twenty, increases to a maximum of 3221 at age seventy-five and then declines to one death at age 104. For women there are only 80 deaths at age twenty, the maximum is at age eighty, where we have 3870 deaths, and the curve has a final value of one death at age 109. It is to be noted that the age at which deaths reach a maximum is five years older for women than for men—eighty as compared with seventy-five. The curve for women also continues for five years more than the curve for men—the highest age being 109 as compared with 104. As a mortality table is something of an artificial construction, however, we should not attach too much importance to the upper limit of the table. I do not suppose that anyone will care to propound a maximum age beyond which human beings cannot survive. For practical purposes, however, there are very few authenticated cases of men living beyond 104 or women beyond 109. I stress the word ' authenticated ' because every so often publicity is given to some individual, or group of individuals, alleged to have attained a spectacular age such as 120. Not surprisingly, such cases are never supported by reliable evidence.

It is natural to ask what effect the decline in mortality has had both on rates of mortality and also on the curve of deaths. The first point that needs to be emphasized is that mortality rates have not been falling uniformly at all ages. The greatest proportionate improvements have been at the younger ages, where in former times a heavy toll was taken by lethal infectious diseases which in the Western world have now been largely eliminated. In middle life, more particularly among men, progress has been slower because these are the age-groups when cancer and ischaemic heart disease account for a substantial proportion of the total deaths and both these diseases have so far been resistant to efforts to overcome them. Indeed, as Professor Morris and Dr Fletcher have shown, the death rates among men from lung cancer and from disease of the coronary arteries have been increasing in recent times. All the same, if we look back over a hundred years, there has been quite a considerable fall in death rates in middle and later life until we come to the really old ages, say over eighty for men or over eighty-five for women. At those late ages we find that death rates have remained stationary for as long as we have reliable records.

The outcome of these trends can be very simply expressed. More people are surviving into old age, but there is as yet no sign of any extension of the limit of human life. I say ' as yet ' because we have all read of experiments to extend life beyond what we at present regard as its natural span. None of us can be dogmatic about the future, and possibly a day may come when living to age 150 will be common. With the world population now doubling itself every thirty-five years or so, I cannot say I would regard this development as much of a blessing. At this moment, however, I am concerned only with the past; and for practical purposes the limits I have already mentioned—104 for men and 109 for women—will serve as terminating points for human life.

The combined effect of falling rates of mortality at younger and middle ages and of stationary rates at old ages is to hollow out the curve of q_x and make it steeper, whereas the effect on the curve of deaths is to shift the maximum point (or mode) up to a higher age, at the same time giving it a more pointed shape. From statistical records we can trace the changes in these curves over the past century or so and measure the improvements that have occurred. But it is important for many purposes to make estimates of the future population; and to do this it is essential to make assumptions about future rates of mortality. We can apply actuarial techniques to estimate what would happen if a particular cause of death were eliminated, or had its incidence reduced. If we were to pursue this process to the limit, we should be left only with deaths due to old age, and the curve of deaths would probably approximate to the familiar Gaussian or Normal distribution. Of course, this is not likely to be reached in practice because there will always be accidents, and there will always be congenital defects causing early death. But it is not without interest to consider for a moment this ' final ' form of the d_x curve, i.e. ignoring the possibility of extension to the human life span. As I suggested in a paper [2] submitted to the Institute of Actuaries Centenary Assembly in 1948, this could be done by eliminating deaths from all causes other than degenerative, and calculating mortality rates for degenerative causes alone. From these rates a life table could be constructed, in which the d_x column would give the ultimate form of the curve of deaths, i.e. the curve we should expect if all deaths were in fact the natural outcome of the process of ageing and none were caused by a supervening disease or accident. Alternatively, it could be regarded as the statistical distribution of the natural life spans of a particular population, where the natural life span is unique for each individual. Different populations would show different distributions and could be compared with one another in much the same way as we compare the relative distribution of the various blood groups.

Although mortality rates are currently falling throughout the world, there are substantial differences between the mortality levels of different countries, some of which are far from easy to explain. Why, for instance, should Norway have the lightest mortality in Europe while Scotland has one of the heaviest ? One would have expected that both racially and climatically Norway and Scotland were sufficiently similar to show a similar mortality experience. Clearly other factors must be involved, one of which may well be the concentration of a substantial proportion of Scotland's population in what is nowadays called the Clydeside conurbation.

Population density is undoubtedly one factor in mortality variations, while another is social and economic status. It is not easy to disentangle these factors as they tend to be closely interrelated. The poorer social classes tend to live in the densely populated areas, so that it is not really possible to say whether their bad mortality experience is due to the density factor or to the economic factor. However that may be, the Registrar General analyses mortality according to five social classes, and as a result we know that there is a marked class gradient in mortality rates, Social Class I, defined as professional and administrative, experiencing the lightest mortality and Social Class V, defined as unskilled labourers, experiencing the heaviest.

Thus we have national differentials in mortality and we have social differentials. To what extent can it be said that there are differentials in the rate of ageing as between one nation and another, or as between one class and another ? I do not want in any way to trespass on Dr Hollingsworth's subject, but I have to affirm here that there is in my view little doubt that longevity is to some degree controlled by heredity ; I think that this was demonstrated by Raymond and Ruth Pearl [7] in their book *The Ancestry of the Long-Lived*. Thus the genetic constitution of a given group or community will determine, or at least will have some effect upon, what I have called the ultimate form of the curve of deaths for that group. Since the deaths represented by the ultimate curve are those attributed to degenerative causes, it seems reasonable to suppose that the rate of ageing is correlated with the natural life span and will thus also be genetically controlled. This does not deny the possibility that the ageing process may be unnaturally speeded up if an individual is exposed to an exceptionally unfavourable environment. Similarly, there may be ways in which the ageing process may be slowed down—indeed Dr Alex Comfort,[3] in his book *The Process of Ageing*, has described such experiments with animals. But leaving these unnatural situations on one side, it seems to me highly probable that the natural process of ageing is genetically controlled and that it is this ageing process that determines the length of the natural life span. If anyone is inclined to regard this as a statement of the obvious, I would point out that it is by no means a necessary conclusion. It would be quite possible to have a predetermined life span without an ageing process. Conversely, as more than one imaginative writer has realized, one can conceive a society in which old age is protracted for long periods without automatically leading to death.

I have already spoken of mortality differences between various groups of people, notably national groups and social groups. Quite a lot of my own work has been concerned with groups defined by some specific medical impairment such as hypertension, overweight, peptic ulcer, diabetes and so on. Such groups experience heavier rates of mortality than normally healthy people and consequently have a diminished expectation of life. Nevertheless modern methods of treatment have been reducing the mortality of many of these groups—diabetes is an obvious example—and greatly increasing their life expectancy.

Although the acquiring of some form of disability may, and frequently does, increase the lethal hazards to which an individual is exposed, there is no reason to suppose—except perhaps in the case of certain comparatively rare conditions—that the ageing process is speeded up. As medical science progresses and treatment becomes more effective, the mortality level of the impaired groups should approach closer and closer to that of a comparable population of healthy lives, but the rate of ageing will probably remain unaffected.

As I remarked earlier, the continuing decline in mortality has been a major factor in causing the population explosion. It is perhaps not out of place to speculate whether this declining trend, instead of continuing, may in fact go into reverse. Even if the Sahara is irrigated and the Polar ice caps are melted, it is only too obvious that a world population which keeps doubling itself every thirty-five years will sooner or later run out of living space. This is not the place or time to discuss whether competition for living space will produce political pressures that will unleash a nuclear war. But it is by no means fanciful to suggest that an overwhelming increase in population density may itself bring about increased mortality rates. Also there must surely come a time when diet shortages induce deficiency diseases—not to mention actual famine—and similarly cause higher mortality. In other words, so far from moving to the state in which virtually all deaths are due to ageing—a state which I regard as the natural objective for any health programme—we may move backwards to the death rates which were common in the eighteenth and nineteenth centuries. In that case, people will start to die younger again, although no doubt from other causes than those which were responsible for the high death rates of former times.

Eventually, a position of equilibrium may be reached with death rates at a level sufficient to maintain a stationary population. By that time, however, family limitation—which now seems a remote possibility in so many parts of the world—may have become a reality and once again the human race may enter a period of falling death rates. I do not propose to speculate any further; but as it seems to be an automatic presumption on the part of a great many people that death rates are going to continue falling indefinitely, I think it is well to emphasize that the current trend is not irreversible and that a period of rising mortality at some future time is by no means outside the range of possibility.

REFERENCES

1. CASE, R. A. M. 1956. Cohort analysis of mortality rates as a historical or narrative technique. *Brit. J. prev. soc. Med.* 10, 159.
2. CLARKE, R. D. 1948. A bio-actuarial approach to forecasting rates of mortality. *Proc. Centen. Assembly Inst. Actuaries*, Vol. II, p. 12.
3. COMFORT, A. 1961. *The Process of Ageing.* London. Weidenfeld and Nicolson.
4. GOMPERTZ, B. 1825. The Law of Human Mortality. *Phil. Trans.* Part II, 513.
5. GRAUNT, J. 1662. *Natural and Political Observations upon the Bills of Mortality.* London.
6. MAKEHAM, W. M. 1867. On the Law of Mortality. *J. Inst. Actuaries*, 13, 325.
7. PEARL, R. and R. DE W. 1934. *The Ancestry of the Long-Lived.* Baltimore. Johns Hopkins.
8. PERKS, W. 1931. Experiments in Graduation of Mortality Statistics. *J. Inst. Actuaries*, 63, 12.
9. REGISTRAR GENERAL FOR ENGLAND AND WALES. 1957. *Decennial Supplement: Life Tables.* London. H.M.S.O.

GENETIC STUDIES ON LONGEVITY

M. J. HOLLINGSWORTH

St. Bartholomew's Hospital Medical College, London

INTRODUCTION

IT is a truism that old individuals in populations consume nutrients and occupy space that would otherwise be available for their offspring. Natural populations survive because young individuals, better adapted to the ever-changing environment, become parents of the next generation, while the less well adapted individuals, perhaps the majority, die.

The length of adult life is related to the process of reproduction. Natural selection ensures that some individuals in a population do not die before they have reproduced, but it would not be expected to promote the survival of these individuals after the reproductive period unless it conferred some advantage.[24, 40] Thus many animals, e.g. migratory salmon, die after shedding their gametes, and annual and biennial plants after shedding their spores or seeds. Some living organisms, of course, are perennial and produce offspring year after year. The giant redwoods (*Sequoia* sp.) and land-locked salmon are examples, but generally a long post-reproductive life does not occur in natural environments, our own species providing the exception which proves the rule. The human post-reproductive life is long because it was advantageous that it should be. Among our ancestors there was selection for long-lived families, because old people served two socially important functions favouring survival. These were looking after the grandchildren while their parents were working, and serving as a store of accumulated knowledge which was passed on to young individuals through the spoken word. The situation has now changed. In our modern society old people are still useful as child-minders but their function as reservoirs of knowledge has been replaced by the printed word.

The death of multicellular organisms, if not premature, is usually preceded by progressive degenerative changes we call senescence. All the individuals of a population may, however, die early in life before these degenerative ageing changes become apparent,[12] and if this happens before reproduction has taken place the population will become extinct. This premature or non-senescent death may be due to lethal or semilethal heritable abnormalities or, more frequently, to exposure to various environmental stresses such as infectious diseases, starvation, lack of living space, and the action of predators (herbivorous animals in the case of plants).

Some species show no senescent changes, are extremely long-lived and appear to be immortal, any deaths that do occur being due to accidents of one kind or another. The giant redwoods and sea anemones are examples of such living organisms with an indeterminate life span.[10] Unicellular organisms are sometimes said to be immortal, but this is a misconception. The individuals in a population of unicells will die, just as will multicellular organisms, if they do not have a continual exchange of energy and matter with their surroundings. A bacterial population derived by inoculating a sterile culture medium with a single individual will become extinct when either the nutrients are exhausted or waste metabolic products reach a toxic concentration.

Maynard Smith [22] put the question: do the cells of an organism senesce because they are in an old organism, or does an organism senesce because its cells are old ? In other words, is ageing and death a consequence of the cellular differentiation or of intrinsic changes in the cells of the organism ? We are not yet in a position to discriminate in favour of either of these alternatives. Both may be true. For example, metabolic wastes may accumulate in the body, cells may die and not be replaced, or irreversible changes may take place in either the nuclei or cytoplasm of cells.[4, 33] Any of these may result in impairment of function and the subsequent death of the organism.

Until recently it was believed that normal cells could be taken from multicellular organisms and maintained indefinitely in tissue culture, provided subcultures are transferred to fresh media at regular intervals. There is now reason to suspect that these cultures have survived long after the death of the parental organism owing to accidental contamination of the cultures with cells when fresh tissue juice was added. In recent studies, when such experimental errors have been excluded, the clones of animal cells have died out after about fifty generations.[13] The same situation seems to exist in higher plants. Street[35] found that apical root meristems of tomato seedlings and of other species could not be subcultured indefinitely because of the increasing inability of the apical cells to carry out their normal metabolic activities. The ageing of ciliate clones and their rejuvenation after sexual reproduction (conjugation) or endomixis, when a new macronucleus is formed, was first reviewed by Jennings[17] and more recently by Siegel.[32] These studies have shown that where environmental causes can be excluded senescence in clones is due to changes, perhaps the accumulation of errors, in either the nucleus or the cytoplasm. Perhaps in those clones which do not age, e.g. the various asexual amoebae and the HeLa cancer cells, these errors do not accumulate in individuals faster than they can be eliminated by natural selection.

LENGTH OF LIFE AND AGEING
IN *DROSOPHILA SUBOBSCURA*

It is not known how long *Drosophila subobscura* lives in its natural environment but probably it is considerably less than can be achieved in laboratory conditions. The length of adult life is a function of temperature, a rise in temperature reducing longevity in a complex way which is currently being investigated. One would therefore expect a *Drosophila* to live longer in the winter months than in the summer. This may be important if *Drosophila* overwinter as adults—as yet an unsettled question. Laboratory studies have shown that the generation time is about one lunar month at 20°C. One would therefore not expect an individual fly to live as long as this. Indeed, it has been observed that the reproductive life of a female at this temperature is less than three weeks, for many of the eggs she lays after the seventeenth day are unfertilized owing to her having run out of sperm. The effective life of males is much shorter than this, for they ejaculate most of their sperm at the first mating. Although they produce more sperm after a few days it is biologically useless, for females in this species normally mate only once in their lives.[21]

Figure 1 is of some *D. subobscura* life tables obtained in laboratory conditions. It shows those of adult male flies from

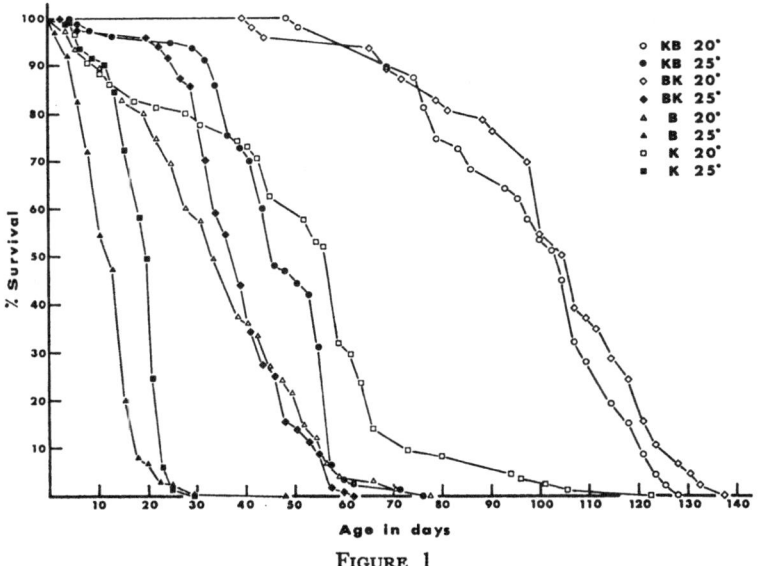

FIGURE 1

Survival curves of **B** and **K** inbred male *Drosophila subobscura* and the reciprocal hybrid males **BK** and **KB** between them (see text) at two environmental temperatures, 20°C and 25°C

two inbred lines, **B** and **K**, and from the reciprocal hybrids between them, **BK** = B♀ × K♂ and **KB** = K♀ × B♂, at two environmental temperatures, 20°C and 25°C.

The mean expectations of life of these flies are given in Table I. Some data are available on the longevity of female flies in this species.[5] They are generally as long-lived as males, but because they are exposed to the additional stress of egg-laying they have not been used in most of our work on ageing.

TABLE I

THE EFFECT OF INBREEDING AND OF TEMPERATURE ON LENGTH OF ADULT LIFE
Mean expectations of life in days and coefficients of variation of inbred
and hybrid male *D. subobscura* at two environmental temperatures.

| | 20°C. | | 25°C. | |
	MEAN	COEFFICIENT OF VARIATION %	MEAN	COEFFICIENT OF VARIATION %
K	51.2 ± 2.7	49.02	18.8 ± 0.3	22.87
B	35.1 ± 2.3	55.61	12.9 ± 0.5	48.06
KB	98.6 ± 2.8	19.48	47.0 ± 1.4	26.62
BK	102.4 ± 3.2	20.83	38.7 ± 0.9	25.97

We will first consider the hybrid life-tables at 20°C. These are basically similar to human life-tables, but there are differences which are important to our understanding of the ageing process. Firstly, the *Drosophila* life-tables are of adult individuals, whereas the human life-table includes individuals that have not completed their development. Soon after emergence from the pupal case, the *Drosophila* imago has completed its development and there are no further cell divisions apart from those in its gonads. For about one third of the life-table the force of mortality is zero. We believe that deaths occurring in this period are due to accidents which are theoretically avoidable. There is no corresponding plateau in the human life-table, because the accidental deaths among young individuals have not been minimized. In constructing *Drosophila* life-tables great care is taken to avoid accidental deaths such as flies escaping during transfer to fresh food vials and others becoming stuck in the moist food medium. We are not always successful in this as Figure 2, some accumulated life-table data, shows. As *Drosophila* age, they become less active and therefore less liable to escape during vial transfer, but they become increasingly liable to stick in the food medium. Although continual improvements in our techniques are being made, we have not yet succeeded in eliminating all stresses from our experimental environment.

The human life-table is usually constructed of individuals of all ages from birth and therefore includes a large number of immature individuals which are especially susceptible to adverse environmental stresses. This is particularly true of human beings under one year of age and is reflected in the dip at the beginning of the life-table. Human infants and young children are exposed to a variety of environmental stresses—infectious diseases, malnutrition, neglect through parental ignorance, and to avoidable environmentally caused

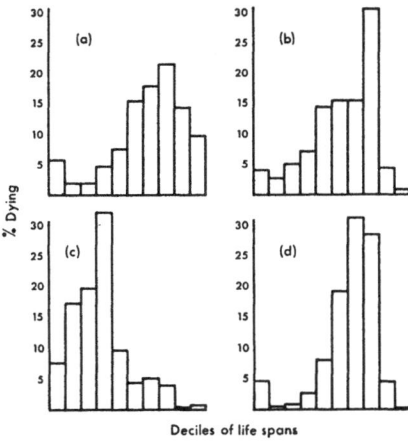

FIGURE 2

Percentage death per decile total life span :

(a) *D. subobscura* hybrid males (**BK** and **KB**). Grouped data of 746 deaths.

(b) **K** inbred male *D. subobscura*. Grouped data of 257 deaths.

(c) **B** inbred male *D. subobscura*. Grouped data of 236 deaths.

(d) Human deaths (England and Wales, 1959). Grouped data of 527,651 deaths.

congenital abnormalities such as the thalidomide syndrome. We would have obtained an analogous initial high mortality in the *Drosophila* life-table if it has been constructed of individuals immediately they had emerged from the pupal cases. Such individuals have unexpanded wings and soft exoskeletons.

They are therefore particularly susceptible to damage in handling in the first hour or two of life. For this reason flies without any visible deformities—such as crumpled wings—are collected from the culture bottles and placed in the food vials in which they are aged, in the afternoon of the day they emerge, the flies emerging in a daily rhythm after dawn.

A second cause of the difference between the *Drosophila* and the human life-tables is due to the genetic constitutions of the two populations from which they are constructed. We believe the hybrid *Drosophila* life-tables to be constructed from genetically more uniform individuals than the human life-tables. The **B** and **K** inbred lines are structurally homozygous and, since they have been maintained by brother-sister matings for more than 160 generations, are possibly largely genetically homozygous at most loci (but see later). The hybrids formed from them show large inversions on three of the four long autosomes, indicating that the inbred lines are homozygous for different orders. When two distinct highly inbred lines are crossed, a population of individuals is obtained with maximally heterozygous genotypes. This has two consequences : *a.* any recessive lethal or semilethal genes in the homozygous condition in one of the inbred parent individuals immediately becomes heterozygous in their offspring and therefore disappear from the phenotype ; *b.* such heterozygous offspring exhibit hybrid vigour, canalised development, and therefore a considerable degree of uniformity.[20]

We see the extreme effects of ' exposed ' lethal and semilethal genes, and of genotype homozygosity and uncanalised development in the phenotypes of the inbred flies from which the hybrids were derived. The **B** inbred line suffers severely from an inbreeding depression, and difficulty is experienced in maintaining it because flies selected as parents frequently die before they have reproduced. The **K** line is quite vigorous by comparison. A measure of the phenotypic variation is given by the coefficients of variation of the flies (Table I). At 20°C the **B** and **K** inbred populations are respectively 2·7 and 2·4 times as variable as the hybrids. However, at 25°C only the **B** population is as variable as at 20°C ; indeed the **K** inbred line is less variable than the hybrids at this temperature. This suggests that in spite of the degree of inbreeding to which it has been subjected the **K** inbred line retains a considerable degree of heterozygosity, perhaps in the form of a balanced polymorphism, and a genotype well adapted to the laboratory conditions.

A third cause of the difference in the shape of the life-tables is the fact that the *Drosophila* life-table is derived from a longitudinal study of a population kept under uniform environmental conditions, whereas the human life-table is a transverse study of a population, the individuals of which have been exposed to varying environmental conditions for different periods of time. The *Drosophila* have been kept in temperature-controlled cabinets, exposed to a twelve-hour daylength and transferred regularly to fresh food vials; but in spite of this care, we are unable to produce flies with identical phenotypes, probably because the inbred lines from which the hybrids are derived are not themselves completely homozygous. If we could obtain flies with identical phenotypes, they would have similar expectations of life and the life-table would become right-angled; as it is, the variation in expectation of life is high. The human life-table changes are greater than those of *Drosophila* largely because of the greater variety of environments to which the individuals comprising the population have been exposed. Owing to this phenotypic variation, we cannot use the life-table to predict the internal state at a given age of any individual organism.

Since deaths occur with increasing frequency as the population ages, changes must have been taking place in the individual organisms while in the plateau phase, i.e. before any deaths due to old age have occurred. The ageing changes are accumulative, but we do not yet know whether these are at a constant or a variable (increasing or decreasing) rate. They result in a progressive decline in the vitality of individuals and an increase in the mortality rate of the population as it ages.

Vitality was defined by Clarke and Maynard Smith[6] as the ability of individuals to withstand the various internal and external stresses which may cause death. In other words, with increase in age, there is a progressive decline in adaptability to environmental stresses, such decline being frequently determined by measuring physiological performance of individuals in an ageing population. Few such studies have been made in ageing *Drosophila*, but it is known that both their tolerance to high lethal temperatures [15] and their fertility [21] are greatly reduced.[15] Bowler and Hollingsworth [3] measured the ability of ageing *Drosophila* to acclimatize to temperature and found that hybrid flies aged at 20°C retained their ability to acclimatize to temperature for about fifty days, but that older flies could not do so. Thus, there is no decline in ability to acclimatize until after the plateau phase of the life-table. We also measured oxygen consumption as a function of age and found that it fell only when the rate of mortality was increasing. This age-dependent decline in oxygen consumption was thought to reflect the lower metabolic activity of the older flies.

Corresponding studies in mammals, including man, have been more extensive. At the present time, however, our knowledge of the ageing process is not sufficient for it to be described in precise mathematical terms; but in spite of this paucity of observational and experimental data there are several mathematical theories of ageing.[36]

THE ROLE OF HEREDITY

The length of life of an organism depends primarily upon its heredity. The evidence for this is overwhelming, the constant difference in mean life span of different species being of major significance in this respect. *Drosophila subobscura* lives about 100 days at 20°C, a mouse about twenty-four months, a man seventy or more years. The variation in life span within a species is also in part due to the various responses of individuals to the internal and external stresses to which they are exposed throughout their lives, adverse responses resulting in a reduction of the life span below the potential maximum. We see this most clearly in the **B** and **K** inbred lines and the hybrids formed from them. The decline in vitality occurs sooner, but not at a faster rate, in the inbreds than in the hybrids. Measuring the expectation of life at 25°C has revealed a significant difference between the **BK** and **KB** males, a difference confirmed in other work.[15] These flies have similar genotypes, apart from their sex chromosomes. The **BK** males receive their X chromosome from their **B** mothers and their Y chromosome from their **K** fathers. The situation is reversed in the **KB** males. The other chromosomes (autosomes) consist of haploid sets from each inbred line in both cases. The Y chromosome is probably not very important in *Drosophila* because it bears few genes, but the X chromosome constitutes about one-fifth of the haploid genotype. It is therefore not surprising that their **B** X chromosome shortens the life of the **BK** males.

Cohen [8] has critically reviewed the past studies on the role of heredity in determining the length of life in our species. She points out that although the many familial (kinship and twin) studies show a correlation between the length of life of related individuals, this may be due to bias involved in the selection, completeness, source or validity of the data used. Thus the familiar saying ' He comes from a long-lived family ', which has been supported by the classical studies, such as those of Beeton and Pearson in 1901 [1] and Pearl and Pearl in 1934 [28] on longevous families, may well be due to environmental effects favouring survival in these families. The nearest approach we have of studies comparable to well-designed animal and plant experiments are those on monozygotic twins, e.g. those of Kallman and his associates. [18] We know at least that the smaller difference between the life spans of monozygotic twins than between those of like-sex dizygotic twins is partly due to the former having identical genotypes.

While lethal and semi-lethal genes can shorten life, I am aware of no evidence in any species that single genes can lengthen life to a similar extent or that there are specific major genes for ageing. This is because length of life is a continuously variable characteristic and would seem to have a genetic basis similar to that of other quantitative traits. Experimental studies of plant and animal populations have lent support to the hypothesis that the genetic basis of continuously variable traits is generally due to the more or less equal and additive effects of genes whose individual genetic effects are less than environmental effects. While we have no direct evidence of a similar mechanism for the determination of human quantitative traits, such as stature, we may safely infer the general validity of the hypothesis.

In animal and plant experiments it has been possible to shift the mean value of quantitative traits by selecting as the parents of the next generation related individuals with extreme phenotypes. It is theoretically possible to do this in man. We could undoubtedly increase human body size in the same way that animal breeders increase egg production in poultry and carcass weight in beef cattle; though one consequence of such selection would probably be the extinction of the population through its members becoming infertile or inviable, a seemingly inevitable consequence of the homozygosity resulting from selection and close inbreeding in a naturally outbreeding population.

Selection for quantitative traits is not always effective. Sometimes it is not possible to shift the mean without the introduction into the genotype of new genetic variation in the form of mutation and recombination. Hollingsworth and Maynard Smith[16] were unable to increase the rate of development in *Drosophila subobscura* by selecting the most rapidly developing individuals in an inbred line as the parents of the next generation, though we had no difficulty in selecting for a reduced rate of development. Analysis of this phenomenon showed that the fastest rate of development occurred in individuals with maximally heterozygous genotypes, the slowest rate of development being in the homozygotes.[23] Fertility and viability are other traits which are maximal in heterozygotes and minimal in homozygotes. This is why inbred lines become infertile and suffer from an inbreeding depression. That longevity is another heterotic trait is suggested by the studies of Comfort,[9] who failed to increase the expectation of life of *D. subobscura* even after eight generations of selection by breeding from the longer-lived individuals. If longevity is due to heterozygosity in man as well as in *Drosophila*, it means there is no possibility of shifting the mean expectation of life much above its present value, and even a small shift will be achieved only by the elimination or minimization of the non-senescent deaths. There will always be a residue of non-senescent deaths from genetic causes.

Although we have very little direct information about the fitness, in the Darwinian sense, of long-lived organisms, it is a reasonable assumption that longevity is a component of fitness. There is the indirect evidence that hybrids, which are fit in the Darwinian sense, are also long-lived. Clarke and Maynard Smith[5] found that the F_1 hybrids between the two inbred lines, **B** and **K**, of *D. subobscura*, lived longer than the parental lines, an observation repeated and extended many times since then. Vetukhiv[39] measured the longevity of hybrids between geographical populations of *D. pseudoobscura* and found they had greater longevities than the parental populations. He also observed that the F_2 generations were inferior to the F_1 and sometimes to the parental strains. The greater longevity of the F_1 generations he attributed to heterosis and the reduced lifespan of the F_2 generations to genetic breakdown in fitness due to gene recombination.

Mourad[25] observed little or no increase in longevity in F_1 hybrids between populations of *D. pseudoobscura* of a common origin, but which had been kept isolated in population cages at different temperatures for several years. He concluded there had been little genetic divergence during their period of isolation and consequently no heterotic effects on crossing them; which is not entirely surprising considering *Drosophila pseudoobscura* kept in population cages are not inbred and maintain a great deal of heterozygosity in their genotypes in the form of balanced polymorphism.

Pearson,[29] in 1895, in considering the curve of deaths of human beings, concluded that the negative skew in the distribution was due to the heterogeneity of the data, which could be considered as a compound of five types of mortality about five different ages. Three of these, mortality in old age, middle life and youth, he regarded as normally distributed, mortality in childhood as markedly positively skewed, while that in infancy as asymptotic.

More recently, Benjamin[2] has reconsidered the human mortality curve. Improvements in social and medical welfare since the time of Pearson's studies have resulted in the elimination of most deaths in childhood and youth. Theoretically we can eliminate all deaths due to accidents and disease, leaving a residue of deaths attributable to genetic causes (lethal and semilethals) and to old age. The question which arises concerns the future shape of the mortality curve. Will it lose its negative skew and become normal? Benjamin implies that it will by distinguishing between senescent and non-senescent deaths, both of which he regards as normally distributed (Figure 3).

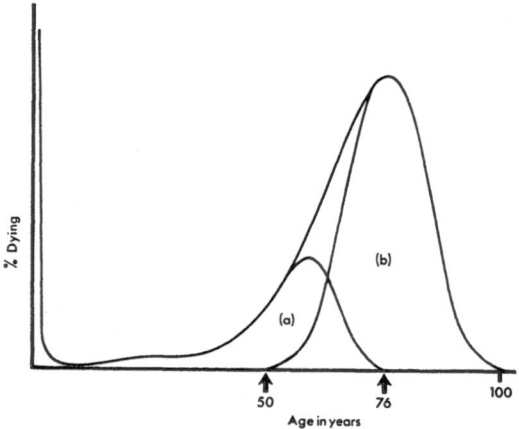

FIGURE 3

English death curve, modified from Benjamin (1959), showing (a) non-senescent deaths, (b) senescent deaths

His hypothesis requires there should be no senescent deaths before age fifty and no accidental deaths after age seventy-six; but one of our criteria of senescent changes is that they should result in an increasing susceptibility to environmental stresses with increase in age. Rotblat, in the discussion of Benjamin's paper, pointed out that non-senescent deaths should properly be distributed over the whole life span. However, when an elderly person is knocked down and killed by a car it is difficult to decide whether it is an accidental or a senescent death. A young person would have been more capable of jumping out of the way, and more likely to recover if he had been hit by the car.

In metric traits due to polygenes acting additively without overdominance, there are substantial parent-offspring correlations, for example, between the heights of fathers and their sons. This seems not to be the case with regard to human longevity. Beeton and Pearson in 1901[1] observed a mean parent-offspring correlation for longevity of 0·1365 and a mean sib-sib correlation of 0·2611, a value twice as great. These correlations are no doubt partly due to environmental causes, but they illustrate an important point made by Haldane in 1949.[11] He pointed out that in so far as fitness is a measurable trait—and here we are assuming longevity to be a measurable component of fitness—it should be substantially correlated in sibs but not at all in parents and offspring (in the absence of environmental factors) in the case of genes where the heterozygotes are fitter than either or both of the homozygotes.

There is thus an alternative possibility to that proposed by Benjamin, namely that the distribution will remain skewed after the minimizing of non-senescent deaths, with the senescent deaths extending well into the younger ages. Figure 2 shows the distributions of length of life for hybrid and inbred *D. subobscura* and for some human data. In each case the frequency of deaths per decile of total life span is illustrated. The human, the hybrid and the **K** inbred *Drosophila* have strongly negatively skewed distributions such as would be expected if length of life were due to genes with over-dominance effects, the longest lived being the most heterozygous individuals and the shorter-lived being the more homozygous individuals in the populations. In marked contrast are the **B** flies, which have a strongly positively skewed distribution such as would be expected if a population were homozygous for the majority of its genes. But the population also contains some comparatively long-lived individuals. It is suggested that these may be rare heterozygotes.

RESEARCH ON THE CAUSES OF AGEING

Whatever the genetic basis of longevity, we can be certain that the degenerative diseases of old age cannot be eliminated. All we can expect in the future is the achievement of the full life span characterestic of our species. While human beings have a life expectancy of seventy plus years, mice live only

P

about two years before they die of the degenerative diseases characteristic of old age in mammals. Rubner [31] noted the inverse correlation between the length of life of animals and their metabolic rate. This concept was developed by Pearl [26] when he postulated that length of life is a function of the genetic constitution of the organism and of the mean metabolic rate of energy expenditure throughout life. This was further developed by Pearl [27] into his *rate of living* hypothesis which postulates:
1. the duration of life varies inversely as the rate of living;
2. different organisms have different *inherent vitalities*, these being part of the gene-controlled patterns of bodily organization;
3. the total vitality of an organism depends on the extent to which matter and energy of exogenous derivation affect the expression of the inherent vitality without altering it. Thus Pearl found that although the mean life span of a starved population of *Drosophila melanogaster* was less than one twenty-fifth that of fed flies, the life-tables, expressed as per cent survival of equivalent life span, were identical (Figure 4).

The length of life of a poikilothermic organism such as *Drosophila* decreases with rise in environmental temperature. Figure 1 shows that at the higher temperature the plateau period is shortened (for the hybrids). Does this indicate that the rates of the ageing processes, which we know do occur in this period, are increased at the higher environmental temperature, as we would expect from Pearl's hypothesis? Clarke and Maynard Smith [6] tested this in two experiments. First, they kept *D. subobscura* adults at a high temperature for about half their expectation of life at that temperature, and then transferred them to a lower temperature. They found that the flies died at the same chronological age as flies kept continuously at the lower temperature; the time spent at the higher temperature did not hasten death. In the second experiment they kept a large population of flies at the lower temperature, and then transferred sample populations at various ages to the higher temperature and found that their mean survival times at the higher temperature declined by about a day for every day they were aged at the lower temperature. This indicated that the rate of ageing is similar at the two temperatures.

When the experiment was repeated on a larger scale by

Hollingsworth in 1966,[14] Clarke and Maynard Smith's observations were fully confirmed. Figure 5 illustrates the results of experiments in which groups of male and female hybrid flies were aged at 20°C and then transferred to 25°C at various ages, their mean survival times at the higher temperature being plotted against age at transfer. There are differences between

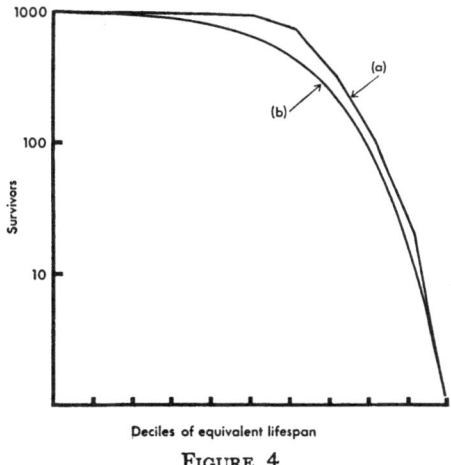

FIGURE 4

Comparison of the survivorship curves of
D. melanogaster males on a centile basis:
(a) starved wild-type males, (b) fed wild-
type males. (Modified from PEARL
(1928))

the survival times of the two types of males and of the males and females which can be attributed to the differences in their genotypes; but in all four cases the survival times at the higher temperature decline by almost exactly a day for every day spent at the lower temperature. This decline in mean survival time continues at a constant rate for about thirty days, after which it is slower. The slope in the first part of the curve is an almost perfect fit to the value of $-1 \cdot 0$ predicted if the rate of ageing is independent of temperature. Pearl's hypothesis predicts that the rate of living (and ageing) should be faster at higher temperatures. This is clearly not true in *Drosophila subobscura.* The rate of ageing is exactly the same at 20°C and at 25°C.

Clarke and Maynard Smith[6] and Maynard Smith[22] have proposed a *threshold* hypothesis of ageing, as an alternative to Pearl's rate of living hypothesis. This distinguishes between the processes of ageing and dying, the former being temperature-independent and the latter temperature-dependent. Their hypothesis can be interpreted as follows. Flies emerge from

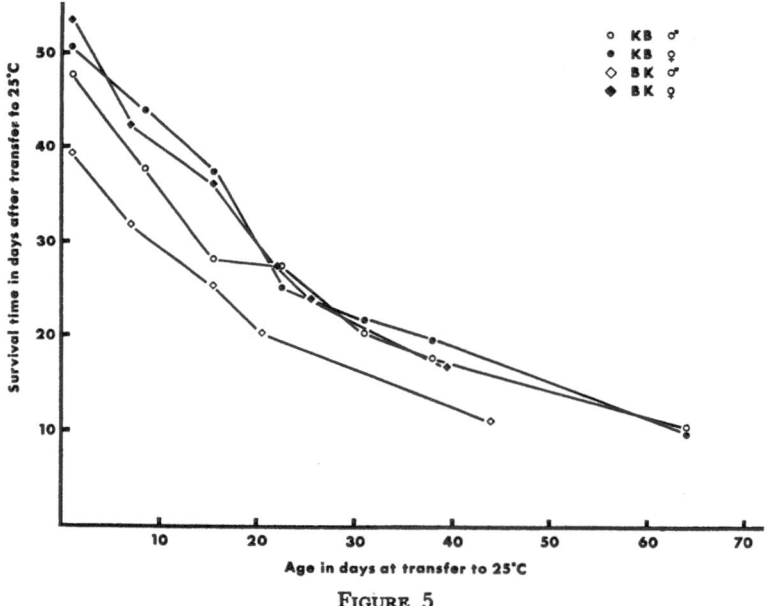

FIGURE 5

The mean survival times in days at 25°C of **BK** and **KB** male and female *D. subobscura* aged at 20°C for various periods before transfer to the higher temperature. (Modified from HOLLINGSWORTH (1966))

their pupal cases, having completed their development, with vitality acquired during the previous developmental stages, the exact amount of vitality depending upon their ability to utilize the available environmental resources. For example, hybrid flies would emerge with more vitality than inbreds, **KB** males with more than **BK** males, and **K** males more than **B** males.

The total amount of vitality runs down as the flies age, but is replenished from time to time when the fly feeds. If a fly feeds on a diet of yeast cells it lives for longer than if on a sucrose diet (Table II). It seems that whereas the sucrose serves only as an exogenous energy source, the yeast diet supplies essential matter for repair and replacement as well, perhaps in the form of amino acids and nucleotides, though this remains to be demonstrated. Tribe [38] has shown a similar dietary effect on longevity in the blowfly, *Calliphora erythrocephala*.

TABLE II

THE EFFECT OF DIET ON LENGTH OF ADULT LIFE
Mean expectations of life (in days) at 20°C of hybrid male
and female *D. subobscura* on yeast and on sucrose diets.

	YEAST	SUCROSE
BK ♂	85·0 ± 7·0	47·4 ± 4·3
KB ♂	87·3 ± 6·6	59·7 ± 3·8
BK ♀	93·1 ± 9·2	51·5 ± 6·1
KB ♀	103·3 ± 9·8	51·4 ± 3·9

The threshold hypothesis postulates that a fly continues to live at a particular temperature so long as the rate of replenishment of its vitality makes up for the rate of utilization, or, in other words, so long as it can maintain its internal organization. If the rate of replenishment of vitality declines with increase in age, a point must be reached at which the fly is forced to draw on its reserves of vitality. It would then enter a state of internal imbalance and begin to die. Maynard Smith [22] postulated that the level of vitality needed by a poikilothermic *Drosophila* to maintain its homeostatic mechanism depends on the temperature at which it is living, more being required at higher temperatures. The individual begins to die as soon as it is exposed to a temperature higher than one at which it can maintain a steady state. It seems, for example, that **BK** males reach this level before **KB** males and **B** males before **K** males because, in both cases, the latter have better homeostatic mechanisms.

We do not yet know what vitality consists of, though some recent work suggests it may be a function of the protein synthesis control mechanism. Clarke and Maynard Smith [7] found that there was a marked and unexpected increase in the amount of protein synthesized in aged *Drosophila* (**BK** males), but it is not yet known which protein(s) this is. Bowler and Hollingsworth[3] interpreted these results in the light of their finding that old flies could not acclimatize to temperature, a process requiring the synthesis of enzymes.[19, 30, 34, 37] We drew two conclusions. First, the proteins made by old flies are not those which permit adaptation to environmental stresses. This would suggest there is a breakdown in the negative feedback control mechanism of protein synthesis, a breakdown which results in positive feedback and a runaway in protein synthesis. Second, old flies make *nonsense* proteins, i.e. molecules with incorrect amino-acid sequences, as a result of faulty transcription or translation of the genetic message.

Current research on ageing in experimental organisms is largely concerned with determining the respective effects on the decline in vitality of accumulated metabolic wastes, loss of cells, and genetic, biochemical and structural changes within cells. *Drosophila* has several advantages as an experimental organism in these respects. First, it ages in an apparently similar way to higher animals. Second, it is an excellent experimental organism. It is easy to breed under strictly controlled environmental conditions, thus making possible the elimination of environmental changes as a cause of ageing. It has a convenient life span. Third, there is no cell replacement in the adult. This facilitates the study of the hypothesis that ageing is related to loss of cells and of cell function. Fourth, our knowledge of the genetics of *Drosophila* far exceeds that of any other group of living organisms. This facilitates the study of the hypothesis that ageing follows a genetically determined programme.[33]

REFERENCES

1. BEETON, M. and PEARSON, K. 1901. On the inheritance of the duration of life and on the intensity of natural selection in man. *Biometrika*, 1, 50.
2. BENJAMIN, B. 1959. Actuarial aspects of human life spans. In *Ciba Fdn. Colloq. Ageing* 5, 2–20. London. Churchill.

3. BOWLER, K. and HOLLINGSWORTH, M. J. 1966. A study of some aspects of the physiology of ageing in *Drosophila subobscura*. *Expl. Geront.* 2, 1–8.

4. CLARK, A. M. 1964. Genetic factors associated with ageing. In *Adv. gerontol. Res.* 1, 207. (Ed. B. L. Strehler.) New York. Academic Press.

5. CLARKE, J. M. and MAYNARD SMITH, J. 1955. The genetics and cytology of *Drosophila subobscura*. XI. Hybrid vigour and longevity. *J. Genet.* 53, 172.

6. CLARKE, J. M. and MAYNARD SMITH, J. 1961. Two phases of ageing in *Drosophila subobscura*. *J. exp. Biol.* 38, 679.

7. CLARKE, J. M. and MAYNARD SMITH, J. 1966. Increase in the rate of protein synthesis with age in *Drosophila subobscura*. *Nature, Lond.* 209, 627.

8. COHEN, B. H. 1964. Family patterns of mortality and life span. *Quart. Rev. Biol.* 39, 130.

9. COMFORT, A. 1953. Absence of a Lansing effect in *Drosophila subobscura*. *Nature, Lond.* 172, 83–84.

10. COMFORT, A. 1964. *Ageing, the Biology of Senescence*. London. Routledge.

11. HALDANE, J. B. S. 1949. Parental and fraternal correlations for fitness. *Ann. Eugen., Lond.* 14, 288.

12. HALDANE, J. B. S. 1953. Some animal life tables. *J. Inst. Actuaries*, 79, 83.

13. HAYFLICK, L. and MOOREHEAD, P. S. 1961. The serial cultivation of human diploid cell. strains. *Expl. Cell Res.* 25, 585.

14. HOLLINGSWORTH, M. J. 1966. Temperature and the rate of ageing in *Drosophila subobscura*. *Expl. Geront.* 1, 259.

15. HOLLINGSWORTH, M. J. and BOWLER, K. 1966. The decline in ability to withstand high temperature with increase in age in *Drosophila subobscura*. *Expl. Geront.* 1, 251.

16. HOLLINGSWORTH, M. J. and MAYNARD SMITH, J. 1955. The effects of inbreeding on rate of development and on fertility in *Drosophila subobscura*. *J. Genet.* 53, 295.

17. JENNINGS, H. S. 1939. Senescence and death in protozoa and invertebrates. In *Problems of Ageing*. (Ed. E. V. Cowdry.) Baltimore. Williams and Wilkins.

18. KALLMANN, F. J. 1957. Twin data on the genetics of ageing. In *Ciba Fdn. Colloq. Ageing* 3, 131. London. Churchill.

19. KANUGO, M. S. and PROSSER, C. L. 1959. Physiological adaptation in the goldfish to cold and warm water. II. Oxygen consumption of liver homogenates, oxygen consumption and oxidative phosphorylations of liver mitochondria. *J. cell. comp. Physiol.* 54, 265.

20. LERNER, I. M. 1954. *Genetic Homeostasis*. Edinburgh. Oliver and Boyd.

21. MAYNARD SMITH, J. 1956. Fertility, mating behaviour and sexual selection in *Drosophila subobscura*. *J. Genet.* 54, 261.

22. MAYNARD SMITH, J. 1962. The causes of ageing. *Proc. roy. Soc.(B)* **157**, 115.

23. MAYNARD SMITH, J. and MAYNARD SMITH, S. 1954. Genetics and cytology of *Drosophila subobscura*. VIII. Heterozygosity, viability and rate of development. *J. Genet.* **52**, 152.

24. MEDAWAR, P. B. 1952. *An Unsolved Problem of Biology*. London. Lewis.

25. MOURAD, A. E. 1965. Genetic divergence in M. Vetukhiv's experimental populations of *Drosophila pseudoobscura*. 2. Longevity. *Genet. Res.* **6**, 139.

26. PEARL, R. 1922. *The Biology of Death*. Philadelphia. Lippincott.

27. PEARL, R. 1928. *The Rate of Living*. London. University of London Press.

28. PEARL, R. and PEARL, R. D. 1934. *The Ancestry of the Long-Lived*. Baltimore. Johns Hopkins.

29. PEARSON, K. 1895. Contributions to the mathematical theory of evolution. II. Skew variation in homogeneous material. *Philos. Trans. (A)* **186**, 343.

30. PRECHT, H., CHRISTOPHERSEN, J. and HENSEL, H. 1952. *Temperatur und Leben*. Berlin. Springer Verlag.

31. RUBNER, M. 1908. *Das Problem der Lebensdauer und seine Beziehungen zum Wachstum und Ernährung*. Munich. Oldenbourg.

32. SIEGEL, R. W. 1967. Control of the life cycle in ciliate protozoa. S.E.B. Symposium No. 21. *Aspects of the Biology of Ageing*. Cambridge Univ. Press.

33. SINEX, F. M. 1966. Biochemistry of Ageing. *Perspect. Biol. Med.* **9**, 208.

34. STANGENBERG, G. 1955. Der Temperatureinfluss auf Lebensprossesse und den Cytochrom c-Gehalt bein Wasserfrosch. *Pflugers Arch. ges. Physiol.*, **260**, 320.

35. STREET, H. E. 1967. Ageing of root meristems. *S.E.B. Symposium No. 21. Aspects of the Biology of Ageing*. Cambridge Univ. Press.

36. STREHLER, B. L. 1962. *Time, Cells and Ageing*. New York and London. Academic Press.

37. SUHRMANN, R. 1955. Weitere Untersuchungen zur Temperaturadaptation der Sauerstoffbindung des Blutes von *Rana esculenta L. Z. vergl. Physiol.* **39**, 507.

38. TRIBE, M. A. 1966. The effect of diet on longevity in *Calliphora erythrocephala Meig. Expl. Geront.* **1**, 269.

39. VETUKHIV, M. 1957. Longevity of hybrids between geographic populations of *Drosophila pseudoobscura*. *Evolution, Lancaster, Pa.* **11**, 348.

40. WILLIAMS, G. C. 1957. Pleiotropy, natural selection and the evolution of senescence. *Evolution, Lancaster, Pa.* **11**, 398.

INDEX OF SUBJECTS

Aberdeen, 29–42
ABO incompatibility, 20–8
Abortion, spontaneous, 12, 14, 43, 45, 56
Adopted children, 41–2
Africa, 149
  Nigeria, 99–110
  South, 66–76
Ageing, 143–212
  causes of, 207–12
  consequences of, 145–61
  genetic vs. environmental factors, 166–8, 178–80
  and housing, 157–60
  and illness, 151, 155–7
  and mortality, 183–93
  psychiatric aspects, 163–80
  unemployment, 151
Air pollution, 82–4
Alcoholism, 130–9
America
  Mexico, 106
  South, 149
  see U.S.A.
American Medical Association, 131
Asia, 149
Australia, 67, 133

Bankers, 68
Bantu, 66–9
BBC, 160–1
Bethlem Royal Hospital, 126
Bills of mortality, 183
Bradford, 23
British Medical Journal, 89
Broken homes: *see* Parental deprivation
Busmen, 68–72

Canada, 67, 114
  Survey in Stirling County, 98–110

Cancer, 62, 63–4, 78–89, 134, 184, 188
' Cape-coloureds ', 66–76
Chromosome abnormalities, 3–17
Cigarettes: *see* Tobacco
Coombs test, 26
Cornell Medical Index Health Questionnaire, 115
Cornell Program in Social Psychiatry 97
Cri-du-Chat syndrome, 3–17

Deafness, 168
Death
  accidental, 46, 63, 189
  adolescent, 61
  age at, 43–56, 60–2, 147–50, 183–93
  from cancer, 78–86
  causes of, 43–56, 59–64
  heart disease, 66–76
  infant, 37–8, 43–56, 60, 189
  pre-school age, 60–1
  rates, 183–93
  school age, 61
  and social class, 190
  stillbirth, 35–8, 43–56
  trends in, 59–64
  see also Suicide
Denmark, suicide rates in, 126
Divorce, 173, 176
Down's syndrome, 3–17
Drosophila, 196–212

Educational achievement, 31
Edwards's syndrome, 3–17
English Life Table No. 11, 183–8
Eskimos, 106
Eugenics Society, v, ix, 29, 139

Family growth, 29–42
Fertility, 148–50, 192–3
France, 106

Gammaglobulin, 25–6
General Register Office, 33, 42
Germany, 23

Headaches, 115–19
Hermaphroditism, 6

Illegitimacy, 126
India, 66
IQ
    of adopted children, 41–2
    and birth weight, 30
    and family size, 39
    and height, 41
    and maternal age, 33–4, 38
    and pregnancy number, 32, 33, 39
    and social class, 31, 38–41
Ischaemic heart disease, 62, 66–76,
    188
Isolation, social, 127, 158, 172, 173–4,
    178

Klinefelter's syndrome, 9, 16

Lancet, 17
Leeds, 23
Liverpool, 20–8
London, 127, 134, 183
Longevity, 188–9, 191, 194–212
Luxembourg, 67

Malformations, 48–52
Maternal age, 15, 17, 33–4, 38
Maudsley Hospital, 126
Medical Research Council,
    Social Psychiatry Research Unit,
    115
    Tristran da Cunhan Study Unit,
    115–19
Mental deficiency, 104
Mental illness: see Psychiatric dis-
    orders
Migration, 146–7
Ministry of Labour and National Ser-
    vice, 162
    Length of Working Life, 152–5

Ministry of Pensions and National
    Insurance, 155, 162
    Circumstances of Retirement Pen-
    sioners, 152–5
    Inquiry into Incapacity for Work,
    112–20
    Reasons given for Retiring, 152–5
Mongolism: see Down's syndrome
Moray House tests, 31–2
Mortality: see Death

National Assistance Board, 173
National Old People's Welfare Coun-
    cil, 161
Negroes, 132
Newcastle Survey, 169
New York
    alcoholism, 132
    State Psychiatric Institute, 123
New Zealand, 67
Norway, 190

Obesity, 72

Parental deprivation, 126
Patau's syndrome, 3–17
Paternal age, 17
Pensioners: see Retirement
Population
    age distribution, 146–50, 163
    and world resources, 192
Pregnancy, multiple, 50
Post-neonatal death: see Death, infant
Pre-Retirement Association, 151, 161
Psychiatric disorders, 97–139
    Canadian and Nigerian surveys,
    97–110
    loneliness, 127, 158, 172, 173–4, 178
    sex of sufferers, 105–7
    and social class, 171–2

Registrar General, 59, 193
Religion
    Baptists, 110
    Protestant, 110
    Roman Catholics, 110
Retirement, 151–5, 160–1

Rh factor, 20–8
Royal College of Physicians, 74, 75–6

Schizophrenia, 163, 166, 167, 168
Scotland, 128, 133, 138, 190
Senility, 163–6
Sheffield, 23
Stillbirth rates: see Death
Suicide, 63, 122–9, 134
Sweden, 67, 106, 179
  suicide rates, 122–3, 128
Switzerland, 67

Thrombosis: see Ischaemic heart disease
Tobacco, 64, 73–5, 78–89
Tristan da Cunhans, 115–19
Turner's syndrome, 3–17
Twins, 85–6, 123–6, 203

U.S.A., 23–5, 67
  Alaska, 106
  alcoholism in, 131–2
  Baltimore, 23
  California, 70, 134
  Connecticut, 134
  retirement, 152, 172
  suicide rates, 128

Venereal disease, 137

Wales
  psychiatric disorders in, 112–21
Widowhood, 151, 172
Witchcraft, 110
World Health Organization, 9, 140

Yoruba: see Africa, Nigeria
Yugoslavia, 66

INDEX OF AUTHORS

Aicardi, J., 5, 18
Alksne, H., 131, 139
Anastasio, M. M., 123–5, 129
Antonis, A., 66, 76
Archambault, L., 5, 18
Archer, M., 76
Aubrey, K. V., 83, 91
Auerbach, O., 79, 89

Backett, E. M., 143–4
Bailey, M. B., 131, 132, 139
Bain, A. D., 6, 8, 17
Bamatter, F., 13, 18
Bandel, 134, 139
Beamish, P., 164, 169, 172, 176, 181
Beatty, R. A., 8, 17
Beeton, M., 203, 207, 212
Benjamin, B., 205–6, 212
Berger, R., 5, 18
Bergmann, K., 164, 171, 174, 181
Berkson, D. M., 67, 77
Berkson, J., 81, 89
Betke, K., 20, 27
Blank, C. E., 6, 17
Blessed, G., 164, 169, 180
Böök, J. A., 13, 17
Bowler, K., 202, 212, 213
Braun, H., 20, 27
Brock, J. F., 66, 76
Bronte-Stewart, B., 66, 76
Bross, I. J., 83, 92
Brothers, C. R. D., 3, 18
Brown, A. C., 114, 121
Brown, D. A., 83, 89
Browne, N. M., 165, 181
Brownlee, K. A., 86, 89
Broyer, M., 5, 18
Brunetti, P. M., 106, 111
Buck, C. W., 114, 121
Buck, S. F., 83, 89
Bühler, E., 13, 18

Campbell, H., 59, 65
Case, R. A. M., 184, 193
Casey, M. D., 6, 17
Chapman, J. M., 67, 71, 76
Chigot, P.-L., 6, 19
Chown, B., 26, 28
Christakis, G., 76
Christophersen, J., 212, 213
Chung, C.-Y., 5, 19
Clark, A. M., 195, 213
Clark, F. le Gros, 154, 162
Clarke, C. A., 20–8
Clarke, J. M., 197, 201, 208–9, 210–212, 213
Clarke, R. D., 155, 183–93
Clemmeson, J., 83, 86, 89
Clow, H. E., 165, 180
Cohen, B. H., 22, 27, 203, 213
Cohen, D. B., 67, 77
Cohen, F., 22, 27
Cohen, J., 79, 89
Comfort, A., 191, 193, 195, 204, 213
Commins, B. T., 83, 84, 90, 91
Cook, J. W., 79, 89
Cook, N. G., 122, 129
Cornfield, J., 81, 89
Cox, P. R., 145–62
Creemers, J., 6, 19
Cruveiller, J., 5, 18

Dahlgren, K. G., 122, 129
Darnell, A., 5, 19
Dawber, T. R., 76
Dean, G., 80, 82, 89
Dencker, S. J., 85, 90
Doll, R., 79, 80, 83, 84, 86, 88, 90, 91
Doust, J. W. L., 114, 121
Downes, J., 114, 121

Edwards, G., 134, 139
Ellis, J. R., 13, 18
England, L., 85, 91

Epstein, F. H., 67, 76
Essen-Moller, E., 106, 111
Eysenck, H. J., 85, 87, 90

Fairbairn, A. S., 83, 84, 90
Falk, H., 83, 90
Fein, H., 76
Ferrier, P., 13, 18
Ferrier, S., 6, 13, 18
Fisher, R. A., 86, 90
Fisher, R. E. W., 84, 90
Fletcher, C. M., 78–92, 188
Fraser, F. C., 12, 19
Freda, V. J., 22, 23, 27, 28
Freudenberg, K., 134, 139
Friberg, L., 85, 90
Friedman, G., 76
Friedman, M., 70, 76

Gallinek, A., 165, 180
Gammon, E. J., 84, 90
Gardner, M. J., 67, 75, 76
Garfinkel, L., 79, 89
Garside, R. F., 164, 169, 172, 180
Gemmell, E., 6, 17
Gibbs, J. P., 128, 129
Glatt, M. M., 134, 139
Godden, J. O., 114, 121
Goldhamer, H., 163, 180
Gompertz, B., 186–7, 193
Gorman, J. G., 22, 23, 27, 28
Graunt, J., 183, 193
Green, J. R., 133, 139
Gropp, A., 13, 18
Grossi-Bianchi, M. L., 6, 19
Grossman, G., 134, 139
Guirão, M., 6, 18
Gunn, W., 84, 90

Haberman, P. W., 131, 132, 139
Haenszel, W., 80, 81, 82, 89, 90
Hahn, W., 70, 77
Haines, M., 6, 18
Haldane, J. B. S., 195, 207, 213
Hall, B., 5, 19
Hall, Y., 67, 77
Halley, E., 183

Hammond, E. C., 79, 81, 82, 89, 92
Harding, J. S., 97, 98, 111
Hawker, A., 134, 139
Hayflick, L., 196, 213
Heady, J. A., 35, 42, 75, 76
Heasman, M. A., 35, 42
Heimann, R. K., 79, 89
Hensel, H., 212, 213
Hensman, C., 134, 139
Hill, A. B., 79, 80, 86, 88, 90
Hill, G. H., 23, 28
Hinkle, L. E., 120, 121
Hollingsworth, M. J., 191, 194–214
Hopkins, B., 166, 180, 181
Hueper, W. C., 83, 90
Hughes, C. C., 97, 98, 104, 111
Hughes, G. O., 84, 90
Hull, A., 20, 28

Illsley, R., 29–42
Inhorn, S. L., 6, 19
Israels, L. G., 20, 26, 28

Jago, G. C., 3, 18
James, G., 76
Jampel, S., 76
Jenkins, C. D., 70, 76
Jennings, E. R., 23, 28
Jennings, H. S., 196, 213
Jérôme, H., 5, 18
Jones, D. L., 78, 90
Jonsson, E., 85, 90
Jussen, A., 13 .18

Kagan, A., 67, 75, 76
Kaij, L., 85, 90
Kallman, F. J., 123–5, 129, 203, 213
Kannel, W. B., 76
Kanugo, M. S., 212, 213
Karvonen, M. J., 77
Kasahara, S., 5, 19
Kay, D. W. K., 164, 166–9, 171, 172,
    174, 176, 179, 180–1
Keller, M., 131, 139
Kessel, N., 130–40
Keys, A., 66, 68, 76
Keys, M. H., 66, 76

Kissen, D. M., 85, 86, 90
Kleihauer, E., 20, 27
Klein, D., 13, 18
Klempman, S., 6, 18
Kneller, O., 78, 91
Koller, S., 79, 90
Kositchek, R., 70, 77
Kotin, P., 83, 90
Kraus, A., 76
Kurihara, M., 78, 91

Lafourcade, J., 5, 6, 18, 19
Lambo, T. A., 97, 104, 109, 111
Langner, T. S., 106, 111
Laughton, K. B., 114, 121
Lawther, P. J., 84, 90
Leighton, A. H., 96, 97–111
Leighton, D. C., 97, 98, 104, 111
Lejeune, J., 5, 6, 18, 19
Lemon, F. R., 83, 92
Lenz, W., 15, 18
Lerner, I. M., 200, 213
LeShan, L., 86, 91
Levine, P., 20, 27
Levinson, M., 67, 77
Lewis, Aubrey, 95–6
Lilienfeld, A. M., 81, 85, 86, 89, 91, 92
Lindberg, H. A., 67, 77
Lindsey, A. J., 79, 91
Lipscomb, W. R., 134, 139
Longaker, W. D., 114, 121
Lord, M., 6, 17
Loudon, J. B., 115, 116, 121
Loveland, D. B., 80, 82, 90
Lowenthal, M. F., 172, 181

McKeown, T., 43–56, 63, 65, 184
MacMillan, A. M., 97, 98, 111
McNamara, P. M., 76
Macklin, D. B., 97, 98, 104, 111
Makeham, W. M., 186, 193
Marr, J. W., 75, 76
Marshall, A. W., 163, 180
Marshall, R., 13, 18
Martin, W. T., 128, 129
Mason, J. I., 85, 92

Massey, F. J., 66, 71, 76
Masterson, J. G., 13, 17
Mayer-Gross, W., 165, 181
Medawar, P. B., 44, 56, 194, 214
Messinger, H. B., 70, 77
Miller, D. E., 131, 139, 140
Miller, J. R., 14, 19
Miller, W., 67, 77
Mills, G. L., 75, 76
Mintz, B., 7, 8, 18
Mittman, O., 79, 91
Mojonnier, L., 67, 77
Montero, E., 6, 18
Moodie, A. D., 66, 76
Moorehead, P. S., 196, 213
Moores, E., 9, 13, 19
Morris, J. N., 63, 65, 66–77, 188
Morrissey, J. D., 166, 181
Mourad, A. E., 205, 214
Mulford, H. A., 131, 139, 140
Mulnard, J. G., 8, 18
Murphy, J. M., 97, 104, 106, 111

Nolan, J. P., 134, 140
Normand, I. C. S., 13, 18

O'Connor, M., 18
Odunjo, F., 13, 18
Overzter, C., 6, 18

Parkes, A. S., v, ix
Parr, D., 137, 140
Parsons, P. L., 169, 181
Pascua, M., 78, 91
Patau, K. A., 6, 19
Pattison, D. G., 67, 75, 76
Payne, W. W., 83, 90
Pearl, R., 191, 193, 203, 208, 209, 214
Pearl, R. de W., 191, 193, 203, 214
Pearson, K., 203, 205, 207, 214
Penrose, L. S., 5, 13, 17, 18
Perks, W., 187, 193
Pike, M. C., 80, 91
Pilkington, T. R. E., 76
Platt, Lord, ix
Poche, R., 79, 91
Polani, P. E., 3–19, 20

Pollack, W., 22, 23, 27, 28
Pollock, J., 26, 28
Post, F., 165, 166, 181
Precht, H., 212, 213
Preisler, O., 23, 28
Prosser, C. L., 212, 213

Raaschou-Nielsen, E., 85, 91
Raffle, P. A. B., 67, 76, 83, 91
Rapoport, R. N., 98, 111
Rawnsley, K., 112–21
Reid, D. D., 83, 84, 90
Reisman, L. E., 5, 19
Réthoré, M.-O., 5, 18
Rinzler, S. H., 76
Robertson, E. E., 165, 181
Roine, P., 77
Rosenman, R. H., 70, 76
Roth, M., 163–82
Rubner, M., 208, 214

Sainsbury, P., 122, 129
Sakabe, H., 84, 91
Salmon, C., 6, 18, 19
Salvatierra, V., 6, 18
Samuels, N., 117
Santesson, B., 13, 17
Sastre, M., 6, 18
Sawicki, E., 83, 90
Scharer, K., 6, 18
Schneider, J., 23, 28
Scott, J. S., 6, 8, 17
Segi, M., 78, 91
Segni, G., 6, 19
Seltzer, C. C., 85, 91
Sheinberg, J., 132, 139
Shimkin, M. B., 81, 89
Siegel, R. W., 196, 214
Simon, K., 114, 121
Sinex, F. M., 195, 212, 213
Sirken, M. G., 80, 82, 90
Smith, D. W., 6, 19
Smith, J. Maynard, 195, 197, 201, 204, 205, 208–9, 210–12, 214
Soltan, H. C., 14, 19
Spicer, C. C., 59–65
Stalder, G., 13, 18

Stamler, J., 67, 77
Stangenberg, G., 212, 213
Stengel, E., 122–9
Stenstedt, A., 167, 182
Stephenson, J., 76
Stocks, P., 80, 82, 83, 91
Stout, A. D., 79, 89
Straus, R., 70, 76
Street, H. E., 196, 214
Strehler, B. L., 202, 214
Suhrmann, R., 212, 213
Sullivan, J. F., 23, 28
Sydenham, 73

Tabor, E. C., 83, 90
Taeuber, K. E., 80, 82, 90
Tarkowski, A. K., 7, 8, 19
Tarrant, M., 85, 90
Tashiro, W., 134, 139
Taylor, A., 4, 9, 13, 19
Therman, E. M., 6, 19
Thieffry, S., 5, 18
Thomas, C. B., 85, 91
Todd, G. F., 85, 92
Tokuhata, G. K., 86, 92
Tomlinson, B. E., 164, 169, 181
Townsend, P., 127, 129, 154–5, 156–160, 162, 173, 175, 182
Tremblay, M., 98, 111
Tribe, M. A., 211, 214
Tunstall, J., 160, 162
Turpeinen, O., 77
Turpin, R., 5, 6, 18, 19
Tyrer, F. H., 84, 90

Uchida, I. A., 14, 19

Van den Berghe, H., 6, 19
Verresen, H., 6, 19
Vetrukhiv, M., 205, 214

Waller, R. E., 81, 83, 84, 92
Walton, H. J., 125–7, 129
Warburton, D., 12, 19
Warkany, J., 6, 19
Wedderburn, D., 154–5, 156–60, 162, 173, 182

Weinstein, E. D., 6, 19
Werthessen, N. T., 70, 77
White, F. D., 20, 28
Williams, G., 134, 139
Williams, G. C., 194, 214
Williamson, J., 169, 182
Wilson, W., 84, 90
Wilton, E., 6, 18
Winslow, G., 76
Wolfbein, S. L., 152, 162
Wolff, E. de, 6, 18

Wolstenholme, G. E. W., 18
Woolf, M., 85, 90
Wright, G., 79, 92
Wurm, M., 70, 76
Wynder, E. L., 79, 81, 82, 83, 89, 92

Young, Q. D., 67, 77

Zipursky, A., 20, 26, 28
Zuelzer, W. W., 22, 27